T0211921

Theory and Applications of Gaussian Quadrature Methods

Synthesis Lectures on Algorithms and Software in Engineering

Editor
Andreas Spanias, *Arizona State University*

Theory and Applications of Gaussian Quadrature Methods
Narayan Kovvali

ISBN: 978-3-031-00389-9   paperback
ISBN: 978-3-031-01517-5   ebook

DOI 10.1007/978-3-031-01517-5

A Publication in the Springer series
SYNTHESIS LECTURES ON ALGORITHMS AND SOFTWARE IN ENGINEERING

Lecture #8
Series Editor: Andreas Spanias, *Arizona State University*
Series ISSN
Synthesis Lectures on Algorithms and Software in Engineering
Print 1938-1727   Electronic 1938-1735

Theory and Applications of Gaussian Quadrature Methods

Narayan Kovvali
Arizona State University

SYNTHESIS LECTURES ON ALGORITHMS AND SOFTWARE IN ENGINEERING #8

ABSTRACT

Gaussian quadrature is a powerful technique for numerical integration that falls under the broad category of spectral methods. The purpose of this work is to provide an introduction to the theory and practice of Gaussian quadrature. We study the approximation theory of trigonometric and orthogonal polynomials and related functions and examine the analytical framework of Gaussian quadrature. We discuss Gaussian quadrature for bandlimited functions, a topic inspired by some recent developments in the analysis of prolate spheroidal wave functions. Algorithms for the computation of the quadrature nodes and weights are described. Several applications of Gaussian quadrature are given, ranging from the evaluation of special functions to pseudospectral methods for solving differential equations. Software realization of select algorithms is provided.

KEYWORDS

numerical integration, Gaussian quadrature, approximation theory, trigonometric and orthogonal polynomials, bandlimited functions, prolate spheroidal wave functions, spectral methods

Contents

viii

CHAPTER 1

Introduction

Quadrature refers to the use of any of a broad class of algorithms for the numerical calculation of the value of a definite integral in one or more dimensions. In this chapter, we introduce the basic numerical integration problem and the general principle behind the design of quadrature algorithms and briefly describe some important families of quadrature rules.

1.1 QUADRATURE

A **quadrature** rule [Atk89] is an approximation of an integral of the form

$$I(f) = \int_\Gamma f(x)\, w(x)\, dx, \tag{1.1}$$

where w is a non-negative 'weight' function, by the N-point weighted summation formula

$$I_N(f) = \sum_{k=0}^{N-1} f(x_k)\, w_k. \tag{1.2}$$

The positions x_k (termed **nodes**) and weighting coefficients w_k (termed **weights**) are chosen such that the integral (1.1) is evaluated *exactly* by the quadrature (1.2), for some class of functions f (usually polynomials of a given degree).

The main idea behind the design of quadrature rules is to find an approximating family $\{f_k(x)\}_{k=0,1,\dots}$ such that

$$\|f - f_k\|_\infty \triangleq \sup_{x \in \Gamma} |f(x) - f_k(x)| \to 0 \ \text{ as } k \to \infty, \tag{1.3}$$

and define

$$I_N(f) = \int_\Gamma f_N(x)\, w(x)\, dx = I(f_N) \tag{1.4}$$

that is easy to evaluate.

The procedure above implies the following error bound on the approximation:

$$|I(f) - I_N(f)| \leq \int_\Gamma |f(x) - f_N(x)|\, |w(x)|\, dx \leq \|f - f_N\|_\infty \cdot \int_\Gamma |w(x)|\, dx. \tag{1.5}$$

Thus, $I_N(f) \to I(f)$ as $N \to \infty$ with a rate of convergence that is at least as rapid as that of f_N to f on Γ.

The domain Γ is often an interval $[a, b] \subseteq \mathbb{R}$. The nodes x_k and weights w_k generally depend on N (implicit in (1.2)) but not on f and are tabulated. Also, the functions f_k need not be constructed explicitly.

1.2   FAMILIES OF QUADRATURE RULES

Let us now consider two important families of quadrature rules, namely **Newton-Cotes formulas** and **Gaussian Quadratures**.

Newton-Cotes formulas are based on using a low-order polynomial approximation of the integrand f on subintervals of decreasing size. The weight function w is generally equal to 1, and the nodes x_k are equispaced. The $(N + 1)$-point Newton-Cotes formula has the property that it exactly integrates polynomials of degree $\leq N$ (N odd) or $\leq N + 1$ (N even). Examples include the **trapezoidal rule** ($N = 1$) and **Simpson's rule** ($N = 2$). Composite versions exist, where the original integral is broken into a sum of integrals defined over partitions of the domain.

The order of accuracy of Newton-Cotes formulas is only **algebraic**, i.e.,

$$|I(f) - I_N(f)| = O(N^{-K}), \quad \text{as } N \to \infty, \tag{1.6}$$

with some fixed K regardless of the smoothness of f. For example, the error in the trapezoidal rule is $O(N^{-2})$ (2nd-order accurate) and that in Simpson's rule is $O(N^{-4})$ (4th-order accurate). Unfortunately, Newton-Cotes formulas do not converge for many functions f. Moreover, due to **Runge's phenomenon**, they converge only in the absence of round-off errors (it is well known that polynomial interpolation on equispaced nodes is an ill-conditioned problem). Newton-Cotes formulas are therefore rarely used with $N \geq 8$ or so.

Gaussian quadratures are a class of methods that typically make use of polynomial approximations of f of increasing degree. The nodes x_k are roots of certain polynomials and are not equispaced but rather tend to cluster near the interval end-points. The nodes are chosen *optimally* so as to maximize the degree of polynomials that the quadrature integrates exactly. In particular, the $(N + 1)$-point Gaussian quadrature exactly integrates polynomials of degree $\leq 2N + 1$. Because of the $N + 1$ extra degrees of freedom in the form of the nodes, the attainable degree is $N + 1$ greater than the Newton-Cotes formulas. An example of a Gaussian quadrature is **Legendre Gaussian quadrature**, for which the weight function w is equal to 1 and the nodes x_k are roots of Legendre polynomials.

The order of accuracy of Gaussian quadratures can be **exponential** (geometric or spectral accuracy), i.e.,

$$|I(f) - I_N(f)| = O(c^{-N}), \quad \text{as } N \to \infty, \tag{1.7}$$

if f is analytic in the neighborhood of $[a, b]$. If $f \in \mathbf{C}^\infty$, then the error decreases faster than any finite power of $1/N$ as $N \to \infty$. If f has $p - 1$ continuous derivatives in $\mathbf{L}^2$ for some $p \geq 0$ and a pth derivative of bounded variation[1], then the error is proportional to $N^{-(p+1)}$ (algebraic convergence).

---

[1]A function f is said to have **bounded variation** if, over the closed interval Γ, there exists an M such that $\sum_{j=1}^{N} |f(x_j) - f(x_{j-1})| \leq M$ for all $x_0 < x_1 < \ldots < x_N \in \Gamma$.

Notably, it is known that Gaussian quadratures converge for any continuous f and are not adversely affected by round-off errors.

1.3   ORGANIZATION

This article is organized as follows. Chapter 2 discusses the approximation theory of trigonometric and orthogonal polynomials and prolate spheroidal wave functions. The analytical framework of Gaussian quadrature is described in Chapter 3, with special coverage of Gaussian quadratures based on prolate spheroidal wave functions. Several applications of Gaussian quadrature are presented in Chapter 4, ranging from the evaluation of special functions to pseudospectral methods for solving differential equations. Appendix A provides Internet links to some useful mathematical software.

CHAPTER 2

Approximating with Polynomials and Related Functions

Polynomials have long been used in the approximation of functions. By the **Weierstrass approximation theorem**, every continuous function defined on an interval $[a, b]$ can be uniformly approximated as accurately as desired by using a polynomial of sufficiently high degree. In this chapter, we review some important results from the approximation theory of trigonometric and orthogonal polynomials and related functions.

2.1 TRIGONOMETRIC AND ORTHOGONAL POLYNOMIALS

Trigonometric and orthogonal polynomials are the eigenfunctions of special cases of the Sturm-Liouville equation

$$\frac{d}{dx}\left[p(x)\frac{dy}{dx}\right] + [q(x) + \lambda w(x)]y = 0, \tag{2.1}$$

defined on some interval $x \in \Gamma$, and form orthogonal and complete basis sets for the Hilbert space $\mathbf{L}^2(\Gamma)$ on this interval with respect to the weight function w [CH53]. Table 2.1 shows specializations of the Sturm-Liouville equation (2.1) for the functions that are of interest here. Note that for the Fourier case the interval $\Gamma = [0, 2\pi)$ is periodic.

	Γ	$p(x)$	$q(x)$	$w(x)$	$y_j(x)$	λ_j
Fourier	$[0, 2\pi)$	1	0	1	e^{ijx}	j^2
Legendre	$[-1, 1]$	$1 - x^2$	0	1	$P_j(x)$	$j(j + 1)$
Chebyshev	$[-1, 1]$	$\sqrt{1 - x^2}$	0	$\frac{1}{\sqrt{1-x^2}}$	$T_j(x)$	j^2
Laguerre	$[0, \infty)$	xe^{-x}	0	e^{-x}	$L_j(x)$	j
Hermite	$(-\infty, \infty)$	e^{-x^2}	0	e^{-x^2}	$H_j(x)$	$2j + 1$

The orthogonal polynomials can be computed either by using the **Gram-Schmidt Orthog-onalization** procedure [TB97] on the monomials $\{1, x, x^2, \ldots\}$ or by the well known recurrence relations [Sze75]:

- Legendre
  $$P_0(x) = 1, P_1(x) = x, (j+1)P_{j+1}(x) = (2j+1)x P_j(x) - j P_{j-1}(x);$$

- Chebyshev
  $$T_0(x) = 1, T_1(x) = x, T_{j+1}(x) = 2x T_j(x) - T_{j-1}(x);$$

- Laguerre
  $$L_0(x) = 1, L_1(x) = 1 - x, (j+1)L_{j+1}(x) = (2j+1-x)L_j(x) - j L_{j-1}(x);$$

- Hermite
  $$H_0(x) = 1, H_1(x) = 2x, H_{j+1}(x) = 2x H_j(x) - 2j H_{j-1}(x).$$

The MATLAB function `legendre_fn.m` in Listing 2.1 shows an example implementation for the Legendre case. Note that the recurrence relation above has been adapted for computing *normalized* Legendre polynomials $\overline{P}_j(x)$. This function additionally computes the first and second derivatives of the normalized Legendre polynomials.

Figure 2.1 shows plots of the Chebyshev polynomials $T_j(x)$ and Legendre polynomials $P_j(x)$, for $j = 0, \ldots, 15$.

An important property of the orthogonal polynomials is that the polynomial $y_j(x)$ has exactly j distinct roots that lie strictly inside Γ. In particular, the j roots of $y_j(x)$ interleave the $j - 1$ roots of $y_{j-1}(x)$, i.e., there is exactly one root of the former between each two adjacent roots of the latter.

A key result concerning approximation of continuous functions using orthogonal polynomials is as follows. For a continuous function $f \in \mathbf{C}^0(\Gamma)$, the degree $\leq N - 1$ polynomial $p_N^*(x)$ that minimizes the weighted 2-norm approximation error

$$\|f - p_N\|_2 \triangleq \left[\int_\Gamma |f(x) - p_N(x)|^2 \, w(x) \, dx \right]^{1/2} \tag{2.2}$$

exists, is unique, and is given by the orthogonal polynomial series

$$p_N^*(x) = \sum_{j=0}^{N-1} \alpha_j \, y_j(x), \tag{2.3}$$

with expansion coefficients

$$\alpha_j = \int_\Gamma f(x) \, y_j(x) \, w(x) \, dx, \ \text{for } j = 0, 1, \ldots, N - 1. \tag{2.4}$$

In other words, orthogonal polynomials provide **Least-squares approximations** to continuous functions.

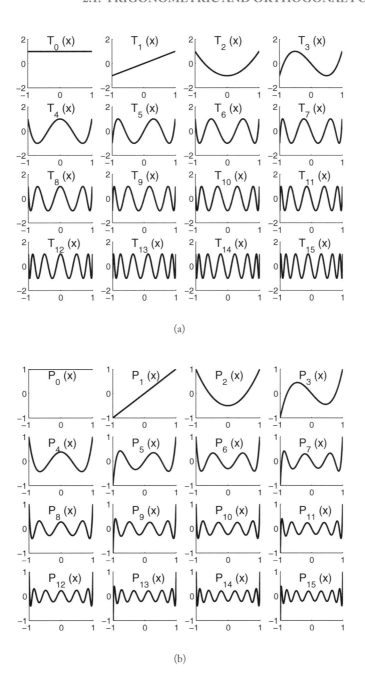

Figure 2.1: (a) Chebyshev polynomials $T_j(x)$ and (b) Legendre polynomials $P_j(x)$, for $j = 0, \dots, 15$.

```
function [P,D1P,D2P] = legendre_fn (x,M)


% Normalized Legendre polynomials \overline{P}_k(x) of order
% k = 0 to M, their first and second derivatives


N = length(x); kvec = [1:M]';
P = zeros(N,M+1); P(:,1:2) = [repmat(1/sqrt(2),N,1) sqrt(3/2)*x];
D1P = zeros(N,M+1); D1P(:,1:2) = [zeros(N,1) repmat(sqrt(3/2),N,1)];
D2P = zeros(N,M+1); D2P(:,1:2) = [zeros(N,1) zeros(N,1)];
v1 = sqrt((2*kvec+1).*(2*kvec-1))./kvec;
v2 = sqrt((2*kvec+1)./(2*kvec-3)).*(kvec-1)./kvec;
for k = 2 : M,
    P(:,k+1) = v1(k)*x.*P(:,k)-v2(k)*P(:,k-1);
    D1P(:,k+1) = v1(k)*(x.*D1P(:,k)+P(:,k))-v2(k)*D1P(:,k-1);
    D2P(:,k+1) = v1(k)*(x.*D2P(:,k)+2*D1P(:,k))-v2(k)*D2P(:,k-1);
end
```

Listing 2.1: `legendre_fn.m`

2.2   PROLATE SPHEROIDAL WAVE FUNCTIONS

Prolate spheroidal wave functions [Bou47, LP61, Sle83, SP61, XRY01] of order zero are the eigen-functions $\psi_j(x)$ of the Helmholtz equation in prolate spheroidal coordinates

$$\left[(x^2 - 1)\frac{d^2}{dx^2} + 2x\frac{d}{dx} + c^2x^2 \right] \psi_j(x) = \chi_j \psi_j(x), \qquad (2.5)$$

where $j = 0, 1, \ldots$ denotes the mode number, χ_j are the associated eigenvalues, and the constant c is known as the **bandwidth parameter**. This equation is also a special case of the Sturm-Liouville equation (2.1), with $\Gamma = [-1, 1]$, $p(x) = 1 - x^2$, $q(x) = -c^2x^2$, $w(x) = 1$, $y_j(x) = \psi_j(x)$, and $\lambda_j = \chi_j$.

Some important properties of the prolate spheroidal wave functions are as follows. For any positive real c,

- the prolate spheroidal wave functions are orthonormal on $[-1, 1]$, with respect to the unit weight function;

- they form a complete basis of $\mathbf{L}^2([-1, 1])$;

- they form a Chebyshev system [CRY99, MRW96, XRY01, YR98] on this interval;

- the wave function $\psi_j(x)$ is symmetric for even j, anti-symmetric for odd j, and has exactly j zeros in $[-1, 1]$;

- when $c = 0$, the prolate spheroidal wave functions are the normalized Legendre functions, i.e., $\psi_j(x)|_{c=0} = \overline{P}_j(x)$.

The best known approach for computing the prolate spheroidal wave functions, due to Bouwkamp [Bou47], is outlined in [XRY01]. The prolate functions are expanded in a series of normalized Legendre polynomials $\overline{P}_k(x)$ as

$$\psi_j(x) = \sum_{k=0}^{M} \beta_{j,k} \, \overline{P}_k(x), \ \ j = 0, \ldots, N - 1. \tag{2.6}$$

The expansion coefficients $\beta_{j,k}$ are computed by solving the symmetric banded eigenvalue problem [XRY01]

$$\mathbf{A}\boldsymbol{\beta}_j = \chi_j \boldsymbol{\beta}_j, \ \ j = 0, \ldots, N - 1, \tag{2.7}$$

where $\mathbf{A}$ is a symmetric banded matrix, with non-zero elements given by

$$A_{k,k} = k(k + 1) + \frac{2k(k + 1) - 1}{(2k + 3)(2k - 1)} c^2, \tag{2.8a}$$

$$A_{k,k+2} = \frac{(k + 2)(k + 1)}{(2k + 3)\sqrt{(2k + 1)(2k + 5)}} c^2, \tag{2.8b}$$

$$A_{k+2,k} = \frac{(k + 2)(k + 1)}{(2k + 3)\sqrt{(2k + 1)(2k + 5)}} c^2. \tag{2.8c}$$

The jth eigenvector $\boldsymbol{\beta}_j$ consists of the expansion coefficients $\beta_{j,k}$, and χ_j is the corresponding eigenvalue. Note that the matrix $\mathbf{A}$ depends on the parameter c, and so do the coefficients $\beta_{j,k}$ and the prolate functions $\psi_j(x)$. The truncation limit $k = M$ for the series in (2.6) is chosen based on the super-algebraic rate of decay of the coefficients $\beta_{j,k}$. Boyd [Boy04] suggests taking M at least equal to $2N$ and testing with a different choice to be certain of the convergence of the expansion coefficients. In [Boy05], the truncation $M = 30 + 2N$ has been shown to be sufficient for convergence up to machine (double) precision.

The MATLAB function `prolate_legendre_expansion.m` shown in Listing 2.2 implements the above algorithm for computing the expansion coefficients $\beta_{j,k}$. The symmetric tridiagonal eigenvalue problem (2.7) can be solved very efficiently using QR iterations [GL96, TB97]; for this, the MATLAB eigensolver `eig` employs the LAPACK [ABB+99] subroutine `DSYEV`.

After the expansion coefficients $\beta_{j,k}$ have been determined, the prolate spheroidal wave functions can be evaluated at any position using (2.6). Observe that the derivatives of the prolate functions can be evaluated in terms of the derivatives of the normalized Legendre polynomials (by differentiating (2.6)).

The MATLAB function `prolate_fn.m` shown in Listing 2.3 uses this method to compute the prolate spheroidal wave functions and their first and second derivatives.

```
function [beta] = prolate_legendre_expansion (c,N)


% Expansion coefficients for prolate spheroidal wave functions
% \psi_j(x), j = 0, ..., N-1 in the basis of normalized Legendre
% polynomials, for bandwidth parameter 0 < c <= c* = (pi/2)*(N+1/2)


M = 2*N+30;        % Truncation limit in normalized Legendre series
k = [0:M]; alpha0 = k.*(k+1)+...
    (2*k.*(k+1)-1)./((2*k+3).*(2*k-1))*c^2;
k = k(1:M-1); alpha2 = ((k+2).*(k+1))./...
    ((2*k+3).*sqrt((2*k+1).*(2*k+5)))*c^2;
A = diag(alpha0)+diag(alpha2,-2)+...
    diag(alpha2,2);               % Symmetric tridiagonal matrix
[V,D] = eig(A); beta = V(:,1:N);  % Expansion coefficients
```

Listing 2.2: `prolate_legendre_expansion.m`

Figure 2.2 shows plots of the prolate spheroidal wave functions $\psi_j(x)$ with $j = 0, 1, \ldots, 15$, for $c = 5$ and 15. Notice the resemblance to the Legendre functions for the smaller value of $c = 5$. The plots also show the increase in oscillations, not only for increasing j, but also for the larger value of $c = 15$.

2.3   APPROXIMATION THEORY

In this section, we recall some basic but important results from the approximation theory of Fourier, Chebyshev, Legendre, and prolate series.

2.3.1   FOURIER SERIES

For any $f \in \mathbf{L}^2([0, 2\pi))$, we can write the Fourier series representation

$$f(x) = \sum_{j=-\infty}^{\infty} \alpha_j \, e^{ijx}, \tag{2.9}$$

where the Fourier coefficients α_j are given by

$$\alpha_j = \frac{1}{2\pi} \int_0^{2\pi} f(x) \, e^{-ijx} \, dx, \ \text{ for } j = 0, \pm 1, \pm 2, \ldots. \tag{2.10}$$

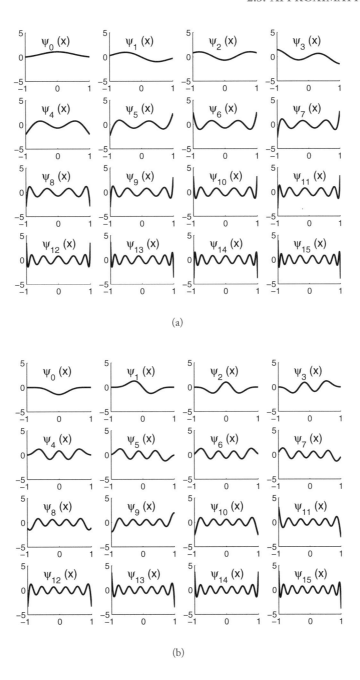

Figure 2.2: Prolate spheroidal wave functions $\psi_j(x)$ ($j = 0, 1, \ldots, 15$), for bandwidth parameter (a) $c = 5$ and (b) $c = 15$.

```
function [Psi,D1Psi,D2Psi] = prolate_fn (x,c,N)


% Prolate spheroidal wave functions \psi_j(x) of order j = 0 to
% N-1, their first and second derivatives, for bandwidth parameter
% 0 < c <= c* = (pi/2)*(N+1/2)


% Expansion coefficients in Legendre basis
[beta] = prolate_legendre_expansion (c,N);
% Normalized Legendre polynomials and their derivatives
M = 2*N+30; [P,D1P,D2P] = legendre_fn (x,M);
% Prolate spheroidal wave functions and their derivatives
Psi = P*beta; D1Psi = D1P*beta; D2Psi = D2P*beta;
```

Listing 2.3: `prolate_fn.m`

We are interested here in the convergence of the N-term truncated Fourier series expansion

$$f^N(x) = \sum_{j=-N/2}^{N/2-1} \alpha_j \, e^{ijx}. \qquad (2.11)$$

For $f \in \mathbf{L}^2([0, 2\pi))$, convergence holds in the $\mathbf{L}^2$ sense:

$$\|f - f^N\|_2 \to 0, \ \text{as } N \to \infty. \qquad (2.12)$$

But $\mathbf{L}^2$-convergence does not imply pointwise convergence of f^N to f at all points of $[0, 2\pi)$. If f is continuous, periodic[1], and of bounded variation on $[0, 2\pi)$, then the convergence is uniform (and hence also pointwise):

$$\max_{x \in [0,2\pi)} |f(x) - f^N(x)| \to 0, \ \text{as } N \to \infty. \qquad (2.13)$$

If f is of bounded variation on $[0, 2\pi)$, then f^N converges pointwise to $(f(x^+) + f(x^-))/2$ for any $x \in [0, 2\pi)$. From Parseval's identity, we have

$$\|f - f^N\|_2 = \left(2\pi \sum_{\substack{j < -N/2 \\ j \geq N/2}} |\alpha_j|^2 \right)^{1/2}, \qquad (2.14)$$

---

[1]A function f in $[0, 2\pi)$ will be called **periodic** if $f(0^+)$ and $f(2\pi^-)$ exist and are equal.

i.e., the size of the truncation error depends on how fast the Fourier coefficients decay to zero.

The following theorem relates the decay of the Fourier coefficients to the smoothness of f.

Theorem 2.1   Decay of Fourier Series Coefficients   If $f \in \mathbf{L}^2([0, 2\pi))$ is m-times continuously differentiable in $[0, 2\pi)$ $(m \geq 1)$ and if $f^{(k)}$ is periodic for all $k \leq m - 2$, then the expansion coefficients in the Fourier series (2.9) satisfy

$$\alpha_j = O(|j|^{-m}), \quad j = \pm 1, \pm 2, \ldots . \tag{2.15}$$

Proof. See [CHQZ88, GO77].   □

In the neighborhood of a point of discontinuity of f or its derivatives, oscillatory behavior of f^N called **Gibbs phenomenon** [GS97, GSSV92] is observed.

As a corollary to Theorem 2.1, the jth Fourier coefficient of a function f which is infinitely differentiable and periodic with all its derivatives on $[0, 2\pi)$ decays faster than any negative power of j. Consequently, the N-term truncated Fourier series (2.11) converges to $f(x)$ more rapidly than any finite power of $1/N$ as $N \to \infty$ for all x.

2.3.2   CHEBYSHEV SERIES

Any function $f \in \mathbf{L}^2([-1, 1])$ can be represented in the Chebyhsev series

$$f(x) = \sum_{j=0}^{\infty} \alpha_j T_j(x), \tag{2.16}$$

where the Chebyshev coefficients α_j are given by

$$\alpha_j = \frac{2}{\pi c_j} \int_{-1}^{1} \frac{f(x) T_j(x)}{\sqrt{1 - x^2}} \, dx, \quad \text{for } j = 0, 1, \ldots, \tag{2.17}$$

with

$$c_j = \begin{cases} 2, & j = 0, \\ 1, & j > 0. \end{cases} \tag{2.18}$$

The Chebyshev polynomials T_j have an alternate form

$$T_j(x) = \cos(j \cos^{-1} x). \tag{2.19}$$

The mapping $x = \cos(\theta)$ shows the equivalence of the Chebyshev series and the Fourier cosine expansion

$$f(\cos(\theta)) = \sum_{j=0}^{\infty} \alpha_j \cos(j\theta). \tag{2.20}$$

The expansion coefficients of f in a Chebyshev series are therefore *identical* to the Fourier cosine coefficients of $f(\cos(\theta))$. Note that while $f(x)$ can be a general (non-periodic) function on $x \in [-1, 1]$, $f(\cos(\theta))$ is always periodic on $\theta \in [0, 2\pi)$.

As for the Fourier series, if the function f is infinitely differentiable, the N-term truncated Chebyshev series

$$f^N(x) = \sum_{j=0}^{N-1} \alpha_j \, T_j(x) \tag{2.21}$$

converges to $f(x)$ faster than any finite power of $1/N$ as $N \to \infty$ for all x.

We are particularly interested in expansions of **bandlimited** functions, which are defined formally as follows [XRY01].

Definition 2.2   Bandlimited Function   A function $f : \mathbb{R} \mapsto \mathbb{R}$ is said to be bandlimited, with bandwidth k, if there exists a positive real number k and a function $\sigma \in \mathbf{L}^2([-1, 1])$ such that

$$f(x) = \int_{-1}^{1} \sigma(t) \, e^{ikxt} \, dt. \tag{2.22}$$

The following theorem quantifies the resolution requirement of the Chebyshev series in the representation of bandlimited functions.

Theorem 2.3   An N-term truncated Chebyshev series representation of a bandlimited function of bandwidth k converges exponentially fast when $N > k$ terms are included. Equivalently, Chebyshev expansions of bandlimited functions converge exponentially when at least π samples are retained per wavelength distance.

Proof. Consider the identity

$$\sin k(x + \tau) = \sum_{j=0}^{\infty} \frac{2}{c_j} J_j(k) \sin\left(\frac{j\pi}{2} + k\tau\right) T_j(x), \ \ x \in [-1, 1], \tag{2.23}$$

where J denotes the Bessel function of the first kind. Since $J_j(k) \to 0$ exponentially fast as j increases beyond k, it follows that an N-term truncated version of (2.23) starts converging very rapidly when $N > k$ terms are included.

Since $\sin k(x + \tau)$ has k/π complete wavelengths lying within the interval $x \in [-1, 1]$, an equivalent way of stating the resolution requirement of Chebyshev series is that Chebyshev expansions of bandlimited functions converge exponentially when at least π samples are retained per wavelength distance. $\qquad \square$

2.3.3   LEGENDRE SERIES

Any function $f \in \mathbf{L}^2([-1, 1])$ has a Legendre series representation

$$f(x) = \sum_{j=0}^{\infty} \alpha_j \, P_j(x), \tag{2.24}$$

where the Legendre coefficients α_j are given by

$$\alpha_j = (j + 1/2) \int_{-1}^{1} f(x) \, P_j(x) \, dx, \ \text{ for } j = 0, 1, \ldots. \tag{2.25}$$

Like the Chebyshev series, the N-term truncated Legendre series

$$f^N(x) = \sum_{j=0}^{N-1} \alpha_j \, P_j(x) \tag{2.26}$$

of an infinitely differentiable function f converges to $f(x)$ faster than any finite power of $1/N$ as $N \to \infty$ for all x.

The following theorem quantifies the resolution requirement of the Legendre series in the representation of bandlimited functions.

Theorem 2.4   An N-term truncated Legendre series representation of a bandlimited function of bandwidth k converges exponentially fast when $N > k$ terms are included. Equivalently, Legendre expansions of bandlimited functions converge exponentially when at least π samples are retained per wavelength distance.

Proof. The Legendre series analog of (2.23) is

$$\sin k(x + \tau) = \sum_{j=0}^{\infty} \sqrt{\frac{\pi}{2k}} \, (2j + 1) \, J_{j+1/2}(k) \, \sin\left(\frac{j\pi}{2} + k\tau\right) P_j(x), \ x \in [-1, 1], \tag{2.27}$$

from which it is clear that an N-term truncated version of (2.27) starts converging exponentially when $N > k$ terms are included.

Therefore, like Chebyshev expansions, Legendre expansions of bandlimited functions converge exponentially when at least π samples are retained per wavelength distance. $\square$

2.3.4   PROLATE SERIES

Any function $f \in \mathbf{L}^2([-1, 1])$ has a prolate series representation

$$f(x) = \sum_{j=0}^{\infty} \alpha_j \, \psi_j(x), \tag{2.28}$$

where the prolate coefficients α_j are given by

$$\alpha_j = \int_{-1}^{1} f(x)\,\psi_j(x)\,dx, \quad \text{for } j = 0, 1, \ldots. \tag{2.29}$$

Like the orthogonal polynomial series, the rate of convergence of an N-term truncated prolate expansion

$$f^N(x) = \sum_{j=0}^{N-1} \alpha_j\,\psi_j(x) \tag{2.30}$$

depends on the smoothness of the function f being expanded. This is because the prolate spheroidal wave functions are also eigenfunctions of a singular Sturm-Liouville problem ($p(x) = 0$ at the boundaries).

Prolate spheroidal wave functions are the natural tool for representing bandlimited functions on an interval. The resolution requirement of the prolate series in the representation of bandlimited functions is quantified by the following theorem [Boy04].

Theorem 2.5   An N-term truncated prolate series representation of a bandlimited function of bandwidth k converges exponentially fast when the bandwidth parameter is chosen as $c = k$ and $N > (2/\pi)k$ terms are included. Equivalently, prolate expansions of bandlimited functions converge exponentially when at least two samples are retained per wavelength distance.

Proof. Consider the operator $F_c : \mathbf{L}^2([-1, 1]) \mapsto \mathbf{L}^2([-1, 1])$ defined by

$$F_c(\varphi)(x) = \int_{-1}^{1} \varphi(t)\,e^{icxt}\,dt. \tag{2.31}$$

The prolate spheroidal wave functions ψ_j are the eigenfunctions of F_c, as shown by

$$\lambda_j\,\psi_j(x) = \int_{-1}^{1} \psi_j(t)\,e^{icxt}\,dt, \quad \text{for all } x \in [-1, 1], \tag{2.32}$$

with λ_j denoting the corresponding eigenvalues. This, combined with orthonormality and completeness of the prolate basis on $\mathbf{L}^2([-1, 1])$, gives

$$e^{icxt} = \sum_{j=0}^{\infty} \lambda_j\,\psi_j(x)\,\psi_j(t), \quad \text{for all } x, t \in [-1, 1]. \tag{2.33}$$

It is known [Boy04] that $\lambda_j \sim \sqrt{2\pi/c}$ for $j < (2/\pi)c$ but falls exponentially for larger j, and so the N-term truncated version of (2.33) converges rapidly when $N > (2/\pi)c$ terms are included. From this, it follows that an N-term truncated prolate series representation of a bandlimited function of bandwidth k converges exponentially when the bandwidth parameter is chosen as $c = k$ and $N > (2/\pi)k$ terms are included.

In contrast to Chebyshev and Legendre expansions of bandlimited functions, which require at least π samples per wavelength for exponential convergence, prolate expansions of bandlimited functions converge exponentially when at least two samples are retained per wavelength distance. □

Remark 2.6   The statement of Theorem 2.5 also shows that in the prolate representation of a bandlimited function of bandwidth k, the optimum choice for the bandwidth parameter is $c = k$ (since this choice of c achieves the Nyquist limit of two samples a wavelength).

Remark 2.7   Chebyshev and Legendre series are the optimal *polynomial* expansions in the $\mathbf{L}^\infty$ and $\mathbf{L}^2$ norms, respectively.

As an example, Figure 2.3 shows plots of the exponential decay of error in the representation of the bandlimited function $\sin kx$ (for $k = 10, 15, 20$, and 25) on $x \in [-1, 1]$, using the Chebyshev, Legendre, and prolate series. We see that the Chebyshev and Legendre series yield exponentially accurate representations when at least π samples are used per wavelength. In contrast, for the representation using the prolate series (with bandwidth parameter $c = k$), exponential accuracy begins after only two samples per wavelength. The plots also show that in all cases these limits are achieved only asymptotically as the wavenumber k increases.

The bandwidth parameter c plays an important role in the approximation theory of the prolate spheroidal wave functions, and its proper choice for a wide variety of function classes is an important issue. Figure 2.4 examines the effect of different choices of the bandwidth parameter c on the representation error for the bandlimited function $\sin 10x$ on $x \in [-1, 1]$, using N terms of the prolate series (for $N = 10, 15, 20$, and 25). The bandwidth of this function is $k = 10$. From the plots, we see that the choice $c \approx k = 10$ is clearly optimum (the representation error is minimum near this value) for all N. For a bandlimited function of bandwidth k, the value $c = k$ usually seems to be the best choice for the bandwidth parameter. In the same plot, for comparison of the errors, we also show the representation error using the Chebyshev polynomial series (which is only a function of N). We see that for the best choices of c, the representation error from the prolate series is *orders of magnitude less* than that from the Chebyshev polynomial series using the same number of terms. Recall also that the error from the prolate series for $c \to 0$ corresponds to the error from the Legendre polynomial series.

But what about representation of functions which are *not* bandlimited? It turns out that as long as the function $f(x)$ is sufficiently smooth, spectral accuracy can be achieved in the representation. This applies to all the basis functions: Chebyshev, Legendre, and prolate. However, for prolate basis functions, the situation is complicated by the fact that the optimum choice of c is usually not fixed but can rather be a function of N. This is illustrated in Figure 2.5, which shows the effect of different

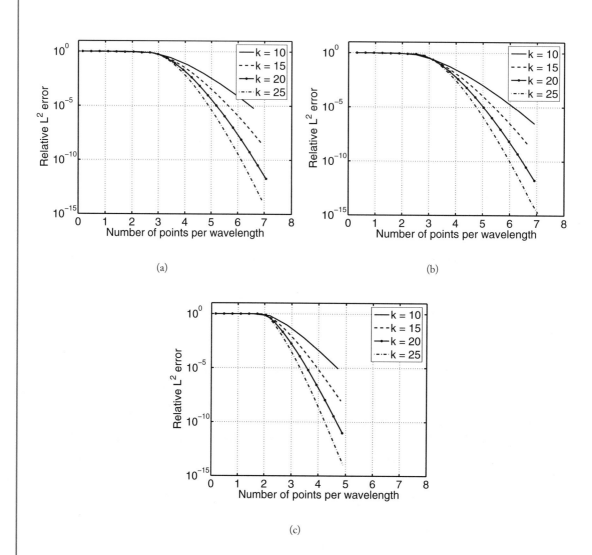

Figure 2.3: Relative error in the representation of the function $\sin kx$ (for $k = 10$, 15, 20, and 25) on $x \in [-1, 1]$, using (a) the Chebyshev polynomial series, (b) the Legendre polynomial series, and (c) the prolate series with bandwidth parameter $c = k$.

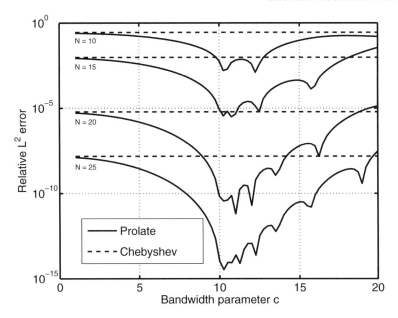

Figure 2.4: Effect of bandwidth parameter c on the representation error for the function $\sin 10x$ on $x \in [-1, 1]$, using N terms of the prolate series (for $N = 10, 15, 20,$ and 25). For comparison, the representation error using the Chebyshev polynomial series (which is only a function of N) is also shown in the same plot.

choices of the parameter c on the representation error for the function $\exp -5x^2$ on $x \in [-1, 1]$, using N terms of the prolate series (for $N = 5, 10, 15, 20,$ and 25). From the plots, we see that the optimum choices of c are not fixed, but increase with N. It is interesting that even though this (Gaussian) function is not strictly bandlimited, for the optimum choices of c, the prolate expansion is still superior to the Chebyshev and Legendre polynomial series.

   Importantly, regardless of whether the function being represented (using an N-term truncated prolate series) is bandlimited or not, all the useful choices of the parameter c must satisfy [Boy04]

$$0 \leq c < c_*(N) = (\pi/2)(N + 1/2). \tag{2.34}$$

If c does not satisfy (2.34), all the prolate functions in the truncated series vanish near the end points of the interval and are therefore incapable of representing arbitrary functions. Empirical convergence studies for representative prototype functions in [Boy04] show that c can be chosen about one-half to two-thirds of $c_*(N)$ for moderate N, approaching the limit as N increases. Guidelines on the suitable choice of c can also be found in [BS05].

   In summary, the plots in Figures 2.3 and 2.4 show that, when at least two samples are used per wavelength, representations of bandlimited functions using prolate spheroidal wave functions are

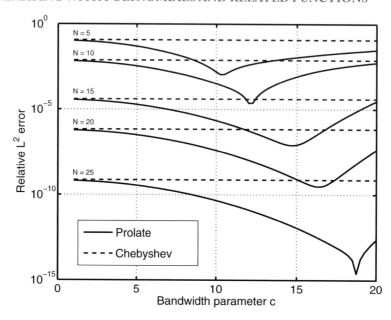

Figure 2.5: Effect of bandwidth parameter c on the representation error for the function $\exp -5x^2$ on $x \in [-1, 1]$, using N terms of the prolate series (for $N = 5, 10, 15, 20,$ and 25). For comparison, the representation error using the Chebyshev polynomial series (which is only a function of N) is also shown in the same plot.

spectrally accurate. Moreover, compared to the Chebyshev and Legendre polynomials, the prolate spheroidal wave functions are better capable of representing bandlimited functions. With proper choice of the bandwidth parameter c, they (a) require a factor of $\pi/2$ samples per wavelength less than the orthogonal polynomials (for achieving a given error tolerance) and (b) yield representations which are orders of magnitude more accurate (using the same number of samples per wavelength).

CHAPTER 3

Gaussian Quadrature

We now discuss the analytical framework of Gaussian quadrature. Below, we make use of the notion of a **Chebyshev system** [CRY99, MRW96, XRY01, YR98], made precise in the following definition.

Definition 3.1 Chebyshev System A sequence of functions $\phi_0, \ldots, \phi_{N-1}$ forms a Chebyshev system on the interval Γ if each of them is continuous and the determinant

$$\begin{vmatrix} \phi_0(x_0) & \ldots & \phi_0(x_{N-1}) \\ \vdots & & \vdots \\ \phi_{N-1}(x_0) & \ldots & \phi_{N-1}(x_{N-1}) \end{vmatrix} \tag{3.1}$$

is nonzero for any sequence of positions $x_0 > \ldots > x_{N-1} \in \Gamma$.

Examples include the set of functions $1, x, x^2, \ldots, x^{N-1}$ on the interval $(-\infty, \infty)$, and the set of functions $1, \cos x, \sin x, \cos 2x, \sin 2x, \ldots, \cos(N-1)x, \sin(N-1)x$ on the interval $[0, 2\pi)$.

Gaussian quadrature is now defined as follows [CRY99, MRW96, XRY01, YR98].

Definition 3.2 Gaussian Quadrature An N-point quadrature formula is a Gaussian quadrature with respect to a set of $2N$ functions $\phi_0, \ldots, \phi_{2N-1}$ on the interval Γ and a weight function $w : \Gamma \mapsto \mathbb{R}^+$ if it integrates exactly each of the $2N$ integrals $\int_\Gamma \phi_j(x) \, w(x) \, dx, j = 0, \ldots, 2N - 1$.

For a set of functions $\phi_0, \ldots, \phi_{2N-1} : \Gamma \mapsto \mathbb{R}$ forming a Chebyshev system on Γ, and a non-negative integrable function $w : \Gamma \mapsto \mathbb{R}^+$, a unique Gaussian quadrature exists for $\phi_0, \ldots, \phi_{2N-1}$ on Γ, with respect to the weight function w. In addition, the weights of this quadrature are positive.

3.1 GAUSSIAN QUADRATURE BASED ON ORTHOGONAL POLYNOMIALS

We start with the most common type of Gaussian quadrature—that based on orthogonal polynomials.

Theorem 3.3 Let $w : \Gamma \mapsto \mathbb{R}^+$ be a weight function, and $\phi_0, \ldots, \phi_{2N-1}$ a sequence of orthogonal polynomials, which are orthonormal with respect to w (in which ϕ_j is of exact degree j). Then the Gaussian quadrature for the functions $\phi_0, \ldots, \phi_{2N-1}$ on Γ, with respect to the weight function w, integrates exactly any polynomial f of degree $2N - 1$ or less. The nodes of this quadrature are the N distinct roots of the polynomial ϕ_N.

Proof. Let f be a polynomial of degree $2N - 1$ or less. Then f can be represented exactly using the basis functions $\phi_0, \dots, \phi_{2N-1}$ as

$$f(x) = \sum_{j=0}^{2N-1} f_j \, \phi_j(x). \tag{3.2}$$

Now, by definition, the nodes and weights of the Gaussian quadrature satisfy the $2N$ equations

$$\int_\Gamma \phi_j(x)\, w(x)\, dx = \sum_{k=0}^{N-1} \phi_j(x_k)\, w_k, \text{ for } j = 0, \dots, 2N-1, \tag{3.3}$$

and we have

$$\begin{aligned}
\int_\Gamma f(x)\, w(x)\, dx &= \sum_{j=0}^{2N-1} f_j \int_\Gamma \phi_j(x)\, w(x)\, dx = \sum_{j=0}^{2N-1} f_j \sum_{k=0}^{N-1} \phi_j(x_k)\, w_k \\
&= \sum_{k=0}^{N-1} \sum_{j=0}^{2N-1} f_j \, \phi_j(x_k)\, w_k = \sum_{k=0}^{N-1} f(x_k)\, w_k,
\end{aligned} \tag{3.4}$$

i.e., the N-point Gaussian quadrature formula is exact for polynomials of degree $2N - 1$ or less.

To show that the nodes of the Gaussian quadrature are the roots of the polynomial ϕ_N, consider the factorization

$$\underbrace{f(x)}_{\deg.\, 2N-1} = \underbrace{q(x)}_{\deg.\, N-1} \underbrace{\phi_N(x)}_{\deg.\, N} + \underbrace{r(x)}_{\deg.\, N-1 \text{ or less}}. \tag{3.5}$$

If we use the fact that q, being a polynomial of degree $N - 1$, can be exactly represented using the basis functions $\phi_0, \dots, \phi_{N-1}$, then

$$\begin{aligned}
\int_\Gamma f(x)\, w(x)\, dx &= \int_\Gamma q(x)\, \phi_N(x)\, w(x)\, dx + \int_\Gamma r(x)\, w(x)\, dx \\
&= \sum_{j=0}^{N-1} q_j \underbrace{\int_\Gamma \phi_j(x)\, \phi_N(x)\, w(x)\, dx}_{=0 \text{ by orthonormality}} + \int_\Gamma r(x)\, w(x)\, dx \\
&= \int_\Gamma r(x)\, w(x)\, dx,
\end{aligned} \tag{3.6}$$

i.e., the integral is independent of q! Now, since f is a polynomial of degree $2N - 1$ or less, the quadrature is exact and

$$\int_\Gamma f(x)\, w(x)\, dx = \sum_{k=0}^{N-1} f(x_k)\, w_k = \sum_{k=0}^{N-1} q(x_k)\, \phi_N(x_k)\, w_k + \sum_{k=0}^{N-1} r(x_k)\, w_k. \tag{3.7}$$

Since the left-hand side is independent of q, so must be the right-hand side, and this is only possible if the nodes x_k of the quadrature are chosen as the N roots of ϕ_N. $\qquad\square$

The following theorem [Atk89] provides an error bound for this quadrature.

Theorem 3.4   Let $w : \Gamma \mapsto \mathbb{R}^+$ be a weight function and $\phi_0, \ldots, \phi_{2N-1}$ a sequence of orthogonal polynomials, which are orthonormal with respect to w (in which ϕ_j is of exact degree j). For any $f \in \mathbf{C}^{2N}(\Gamma)$, the Gaussian quadrature for the functions $\phi_0, \ldots, \phi_{2N-1}$ on Γ, with respect to the weight function w, is such that

$$\left| \int_\Gamma f(x)\, w(x)\, dx - \sum_{k=0}^{N-1} f(x_k)\, w_k \right| \leq \frac{M}{(2N)!} \|\phi_N\|^2, \tag{3.8}$$

where

$$M = \sup_{\xi \in \Gamma} f^{(2N)}(\xi). \tag{3.9}$$

Proof. See [Atk89]. $\qquad\square$

The accuracy of the Gaussian quadrature is dependent on the smoothness of the function f being integrated. In particular, if f is sufficiently smooth, Gaussian quadrature is spectrally accurate [Tre08]. The article [Tre08] provides specific smoothness conditions for f and error bounds showing the spectral accuracy of Gaussian quadrature, and compares it to the closely related Clenshaw-Curtis quadrature [CC60].

3.1.1   COMPUTATION OF THE QUADRATURE NODES AND WEIGHTS

The Gaussian quadrature nodes can be computed by any root-finding algorithm, and the weights w_k can be easily determined by solving the $N \times N$ system of linear equations

$$\int_\Gamma \phi_j(x)\, w(x)\, dx = \sum_{k=0}^{N-1} \phi_j(x_k)\, w_k, \text{ for } j = 0, \ldots, N-1. \tag{3.10}$$

Since polynomials form Chebyshev systems, a unique solution always exists for (3.10). For the Chebyshev case, explicit formulas are known for the Gaussian quadrature nodes and weights:

$$x_j = \cos \frac{(2j+1)\pi}{2N}, \; w_j = \frac{\pi}{N}, \text{ for } j = 0, \ldots, N-1. \tag{3.11}$$

Here we outline a more interesting approach [Gol73, GW69, TB97] based on the three-term recurrence relations satisfied by the orthogonal polynomials.

The three-term recurrent relationship satisfied by orthogonal polynomials is written as:

$$\beta_j \phi_j(x) = (x - \alpha_j)\phi_{j-1}(x) - \beta_{j-1}\phi_{j-2}(x), \text{ for } j = 1, 2, \ldots, N-1;$$

$$\phi_{-1}(x) \triangleq 0, \ \phi_0(x) \equiv 1.$$

(3.12)

Expressing (3.12) in matrix form,

$$x\Phi(x) = \mathbf{J}_N \Phi(x) + \beta_N \phi_N(x)\mathbf{e}_N,$$

(3.13)

where

$$\Phi(x) = [\phi_0(x), \phi_1(x), \ldots, \phi_{N-1}(x)]^T,$$

(3.14)

$$\mathbf{e}_N = [0, 0, \ldots, 1]^T,$$

(3.15)

and

$$\mathbf{J}_N = \begin{bmatrix} \alpha_1 & \beta_1 & & & \\ \beta_1 & \alpha_2 & \beta_2 & & \\ & \cdot & \cdot & \cdot & \\ & & \cdot & \cdot & \cdot \\ & & & \beta_{N-1} & \alpha_N \end{bmatrix}.$$

(3.16)

Solving the eigenvalue problem

$$\mathbf{J}_N \mathbf{v}_j = \lambda_j \mathbf{v}_j, \ j = 0, \ldots, N-1,$$

(3.17)

where

$$\mathbf{v}_j^T \mathbf{v}_j = 1$$

(3.18)

and

$$\mathbf{v}_j^T = [v_{0,j}, v_{1,j}, \ldots, v_{N-1,j}],$$

(3.19)

the nodes and weights are obtained as [Gol73, GW69, TB97]

$$x_j = \lambda_j, \ w_j = (v_{0,j})^2 \times \int_\Gamma w(x)\,dx.$$

(3.20)

The MATLAB function legendre_gauss_quad.m shown in Listing 3.1 implements this algorithm for the Legendre case. Note that the eigenvalue problem (3.17) can be solved very efficiently using QR iterations [GL96, TB97] because, not only is the matrix $\mathbf{J}_N$ tridiagonal, if the orthogonal polynomials are normalized (i.e., they are orthonormal), it is also symmetric. This is effected with the MATLAB eigensolver eig, which employs the LAPACK [ABB+99] subroutine DSYEV for this purpose.

```
function [x,w] = legendre_gauss_quad (N)

% Nodes and weights for N-point Legendre Gaussian quadrature

n = [0:N-2]'; beta = (n+1)./sqrt((2*n+1).*(2*n+3));
J = diag(beta,-1)+diag(beta,1);    % Symmetric tridiagonal matrix
[V,D] = eig(J);                     % Solve eigenvalue problem
x = diag(D); [x,i] = sort(x);       % Gauss quadrature nodes
x = flipud(x); w = 2*V(1,i).^2';    % Gauss quadrature weights
```

Listing 3.1: `legendre_gauss_quad.m`

Tables of the nodes and weights for Legendre, Chebyshev, Laguerre, and Hermite Gaussian quadratures for several N can be found in [AS65].
The output from this program is shown below.

```
>> [x,w] = legendre_gauss_quad (2)
x =
   0.57735026918963
  -0.57735026918963
w =
   1.00000000000000
   1.00000000000000

>> [x,w] = legendre_gauss_quad (3)
x =
   0.77459666924148
                  0
  -0.77459666924148
w =
   0.55555555555556
   0.88888888888889
   0.55555555555556

>> [x,w] = legendre_gauss_quad (4)
x =
   0.86113631159405
   0.33998104358486
  -0.33998104358486
  -0.86113631159405
w =
```

```
0.34785484513745
0.65214515486255
0.65214515486255
0.34785484513745
```

3.2   GAUSSIAN QUADRATURE BASED ON PROLATE SPHEROIDAL WAVE FUNCTIONS

For the integration of bandlimited functions on an interval, the approximation theory of the preceding chapter motivates Gaussian quadrature based on prolate spheroidal wave functions [XRY01].

Theorem 3.5   Let $f : [-1, 1] \mapsto \mathbb{C}$ be a bandlimited function given by the formula

$$f(x) = \int_{-1}^{1} F(t) \, e^{icxt} \, dt, \tag{3.21}$$

where c is a positive real number. Then the Gaussian quadrature for the zero-order prolate spheroidal wave functions $\psi_0, \ldots, \psi_{2N-1}$ on the interval $[-1, 1]$, with respect to the unit weight function, is such that

$$\left| \int_{-1}^{1} f(x) \, dx - \sum_{k=0}^{N-1} f(x_k) \, w_k \right| \le \|F\| \cdot \sum_{j=2N}^{\infty} |\lambda_j| \cdot \|\psi_j\|_{\infty}^2 \cdot \left(2 + \sum_{k=0}^{N-1} |w_k| \right). \tag{3.22}$$

Proof.   Since the prolate spheroidal wave functions $\psi_0, \psi_1, \ldots$ constitute a complete and orthonormal basis in $\mathbf{L}^2([-1, 1])$, we have

$$e^{icxt} = \sum_{j=0}^{\infty} \left(\int_{-1}^{1} e^{icx\tau} \, \psi_j(\tau) \, d\tau \right) \psi_j(t), \tag{3.23}$$

for all $x, t \in [-1, 1]$. Using the fact that (see (2.32))

$$\lambda_j \, \psi_j(x) = \int_{-1}^{1} e^{icxt} \, \psi_j(t) \, dt, \tag{3.24}$$

for all $x \in [-1, 1]$, we get

$$e^{icxt} = \sum_{j=0}^{\infty} \lambda_j \, \psi_j(x) \, \psi_j(t). \tag{3.25}$$

By definition, the quadrature integrates exactly the first $2N$ eigenfunctions, i.e.,

$$\int_{-1}^{1} \psi_j(x) \, dx = \sum_{k=0}^{N-1} \psi_j(x_k) \, w_k, \text{ for } j = 0, \ldots, 2N-1, \tag{3.26}$$

and we have

$$
\int_{-1}^{1} e^{icxt} \, dx - \sum_{k=0}^{N-1} e^{icx_k t} \, w_k
$$

$$
= \int_{-1}^{1} \left(\sum_{j=0}^{\infty} \lambda_j \, \psi_j(x) \, \psi_j(t) \right) dx - \sum_{k=0}^{N-1} \left(\sum_{j=0}^{\infty} \lambda_j \, \psi_j(x_k) \, \psi_j(t) \right) w_k
$$

$$
= \int_{-1}^{1} \left(\sum_{j=2N}^{\infty} \lambda_j \, \psi_j(x) \, \psi_j(t) \right) dx - \sum_{k=0}^{N-1} \left(\sum_{j=2N}^{\infty} \lambda_j \, \psi_j(x_k) \, \psi_j(t) \right) w_k. \tag{3.27}
$$

Finally,

$$
\int_{-1}^{1} f(x) \, dx - \sum_{k=0}^{N-1} f(x_k) \, w_k
$$

$$
= \int_{-1}^{1} F(t) \left[\int_{-1}^{1} \left(\sum_{j=2N}^{\infty} \lambda_j \, \psi_j(x) \, \psi_j(t) \right) dx - \sum_{k=0}^{N-1} \left(\sum_{j=2N}^{\infty} \lambda_j \, \psi_j(x_k) \, \psi_j(t) \right) w_k \right] dt
$$

$$
\leq \|F\| \cdot \sum_{j=2N}^{\infty} |\lambda_j| \cdot \|\psi_j\|_{\infty}^2 \cdot \left(2 + \sum_{k=0}^{N-1} |w_k| \right), \tag{3.28}
$$

proving the result. $\qquad\square$

The nodes and weights of this quadrature can be computed by solving the $2N \times 2N$ system of nonlinear equations

$$
\int_{-1}^{1} \psi_j(x) \, dx = \sqrt{2} \, \beta_{j,0} = \sum_{k=0}^{N-1} \psi_j(x_k) \, w_k, \text{ for } j = 0, \ldots, 2N - 1. \tag{3.29}
$$

An alternate approach to bandlimited function quadrature is to choose the nodes of the quadrature as the N roots of the zero-order prolate spheroidal wave function ψ_N [XRY01].

The following Lemma [XRY01] provides a required intermediate result for the factorization of a bandlimited function.

Lemma 3.6   *Suppose that $\sigma, \varphi : [-1, 1] \mapsto \mathbb{C}$ are a pair of $\mathbf{C}^2$-functions such that $\varphi(-1) \neq 0$, $\varphi(1) \neq 0$, c is a positive real number and the functions f, p are defined by the formulae*

$$
f(x) = \int_{-1}^{1} \sigma(t) \, e^{2icxt} \, dt, \tag{3.30}
$$

$$
p(x) = \int_{-1}^{1} \varphi(t) \, e^{icxt} \, dt. \tag{3.31}
$$

Then there exist two functions $\eta, \xi : [-1, 1] \mapsto \mathbb{C}$ such that

$$f(x) = p(x) \, q(x) + r(x) \tag{3.32}$$

for all $x \in \mathbb{R}$, with the functions $q, r : [-1, 1] \mapsto \mathbb{C}$ defined by the formulae

$$q(x) = \int_{-1}^{1} \eta(t) \, e^{icxt} \, dt, \tag{3.33}$$

$$r(x) = \int_{-1}^{1} \xi(t) \, e^{icxt} \, dt. \tag{3.34}$$

Proof. See [XRY01]. □

Finally, here is the roots quadrature and its error-bound.

Theorem 3.7   Suppose that $x_0, \ldots, x_{N-1} \in [-1, 1]$ are the roots of the zero-order prolate spheroidal wave function ψ_N. Let the numbers $w_0, \ldots, w_{N-1} \in \mathbb{R}$ be such that

$$\int_{-1}^{1} \psi_j(x) \, dx = \sum_{k=0}^{N-1} \psi_j(x_k) \, w_k, \tag{3.35}$$

for all $j = 0, \ldots, N - 1$. Then for any function $f : [-1, 1] \mapsto \mathbb{C}$, which satisfies the conditions of Lemma 3.6,

$$\left| \int_{-1}^{1} f(x) \, dx - \sum_{k=0}^{N-1} f(x_k) \, w_k \right| \le |\lambda_N| \cdot \|\eta\| + \|\xi\| \cdot \sum_{j=N}^{\infty} |\lambda_j| \cdot \|\psi_j\|_{\infty}^2 \cdot \left(2 + \sum_{k=0}^{N-1} |w_k| \right), \tag{3.36}$$

where the functions $\eta, \xi : [-1, 1] \mapsto \mathbb{C}$ are as defined in Lemma 3.6.

Proof. Is on the same lines as before (see [XRY01] for details). □

Remark 3.8   The reader will note an important distinction between the two quadratures above. While both schemes employ prolate spheroidal wave functions corresponding to bandwidth c, the former is designed for functions of bandwidth c whereas the latter is designed for functions of bandwidth $2c$.

The nodes of this quadrature can be computed using any root-finding algorithm, and the weights can be determined by solving the $N \times N$ system of linear equations

$$\int_{-1}^{1} \psi_j(x) \, dx = \sqrt{2} \, \beta_{j,0} = \sum_{k=0}^{N-1} \psi_j(x_k) \, w_k, \text{ for } j = 0, \ldots, N - 1. \tag{3.37}$$

3.3   GAUSS-LOBATTO QUADRATURE

A variation on the Gaussian quadratures above is to include the end-points of the interval[1] $[-1, 1]$ as nodes, which can facilitate the application of boundary conditions in the solution of boundary value problems. When both end-points -1 and 1 are included in the nodes, the Gaussian quadrature is known as a **Gauss-Lobatto quadrature**.

Definition 3.9   Gauss-Lobatto quadrature   An N-point quadrature formula is a Gauss-Lobatto quadrature with respect to a set of $2N - 2$ functions $\phi_0, \ldots, \phi_{2N-3}$ on the bounded interval $[-1, 1]$ and a weight function $w : [-1, 1] \mapsto \mathbb{R}^+$ if it integrates exactly each of the $2N - 2$ integrals $\int_{-1}^{1} \phi_j(x)\, w(x)\, dx$, $j = 0, \ldots, 2N - 3$ and has $x_0 = 1$ and $x_{N-1} = -1$ as its two extremal nodes.

Remark 3.10   Gauss-Lobatto quadrature has two degrees of freedom less—and consequently, slightly lower accuracy—than conventional Gaussian quadrature because two nodes have been restricted to coincide with the interval end-points. Error bounds similar to those of conventional Gaussian quadrature can be derived.

The following result holds for Gauss-Lobatto quadrature based on orthogonal polynomials.

Theorem 3.11   Let $w : [-1, 1] \mapsto \mathbb{R}^+$ be a weight function and $\phi_0, \ldots, \phi_{2N-3}$ a sequence of orthogonal polynomials, which are orthonormal with respect to w (in which ϕ_j is of exact degree j). Then the Gauss-Lobatto quadrature for the functions $\phi_0, \ldots, \phi_{2N-3}$ on $[-1, 1]$, with respect to the weight function w, integrates exactly any polynomial f of degree $2N - 3$ or less. The nodes of this quadrature are the N distinct roots of the polynomial $(1 - x^2)\phi'_{N-1}$.

Proof.   Is similar to that for conventional Gaussian quadrature. $\square$

The Gauss-Lobatto quadrature nodes can be computed by any root-finding algorithm, and the weights w_k can be easily determined by solving the $N \times N$ system of linear equations (3.10). As for conventional Gaussian quadrature, in the Chebyshev case, explicit formulas are known for the Gauss-Lobatto quadrature nodes and weights:

$$x_j = \cos\left(\frac{j\pi}{N-1}\right), \ w_j = \begin{cases} \frac{\pi}{2N-2}, & j = 0, N-1 \\ \frac{\pi}{N-1}, & 1 \leq j \leq N-2 \end{cases}. \tag{3.38}$$

The eigenvalue approach [GW69, Gol73, TB97] also remains applicable. We do not give details but provide a MATLAB implementation `legendre_gauss_lobatto.m` in Listing 3.2 for the Legendre case.

---

[1]The discussion in this part of the chapter pertains only to quadratures on the bounded interval $[-1, 1]$.

```
function [x,w] = legendre_gauss_lobatto (N)

% Nodes and weights for N-point Legendre Gauss-Lobatto quadrature

n = [0:N-3]'; beta = (n+1)./sqrt((2*n+1).*(2*n+3));
J_Nm1 = diag(beta,-1)+diag(beta,1);
gamma = (J_Nm1+eye(N-1))\[zeros(N-2,1); 1];
mu = (J_Nm1-eye(N-1))\[zeros(N-2,1); 1];
beta_Nm1 = sqrt(2/(gamma(N-1)-mu(N-1)));
alpha_N = 1+mu(N-1)*2/(gamma(N-1)-mu(N-1));
Jtilde_N = [J_Nm1 beta_Nm1*[zeros(N-2,1); 1]; beta_Nm1*...
    [zeros(N-2,1); 1]' alpha_N]; % Symmetric tridiagonal matrix
[V,D] = eig(Jtilde_N);           % Solve eigenvalue problem
x = diag(D); [x,i] = sort(x);    % Gauss-Lobatto quadrature nodes
x = flipud(x); w = 2*V(1,i).^2'; % Gauss-Lobatto quadrature weights
```

Listing 3.2: `legendre_gauss_lobatto.m`

The output from this program is shown below.

```
>> [x,w] = legendre_gauss_lobatto (3)
x =
   1.00000000000000
   0.00000000000000
  -1.00000000000000
w =
   0.33333333333333
   1.33333333333333
   0.33333333333333

>> [x,w] = legendre_gauss_lobatto (4)
x =
   1.00000000000000
   0.44721359549996
  -0.44721359549996
  -1.00000000000000
w =
   0.16666666666667
   0.83333333333333
```

```
    0.83333333333333
    0.16666666666667

>> [x,w] = legendre_gauss_lobatto (5)
x =
    1.00000000000000
    0.65465367070798
   -0.00000000000000
   -0.65465367070798
   -1.00000000000000
w =
    0.10000000000000
    0.54444444444444
    0.71111111111111
    0.54444444444444
    0.10000000000000
```

For Gauss-Lobatto quadrature based on prolate spheroidal wave functions, two different choices are popular for the nodes and weights. The first approach is to choose the nodes and weights by solving the $(2N - 2) \times (2N - 2)$ system of nonlinear equations [Boy04, Boy05, CGH05]

$$\int_{-1}^{1} \psi_j(x)\, dx = \sqrt{2}\, \beta_{j,0} = \sum_{k=0}^{N-1} \psi_j(x_k)\, w_k, \text{ for } j = 0, \ldots, 2N - 3, \tag{3.39}$$

where $x_0 = 1$ and $x_{N-1} = -1$. The resulting nodes are termed the **prolate Gauss-Lobatto** nodes. The second approach chooses the nodes as the N roots of the function $(1 - x^2)\psi'_{N-1}$. These are termed **prolate-Lobatto** nodes, and the weights are then computed by solving the $N \times N$ system of linear equations [XRY01, CGH05, KLC05]

$$\int_{-1}^{1} \psi_j(x)\, dx = \sqrt{2}\, \beta_{j,0} = \sum_{k=0}^{N-1} \psi_j(x_k)\, w_k, \text{ for } j = 0, \ldots, N - 1. \tag{3.40}$$

Remark 3.12   Owing to the relationship between prolate spheroidal wave functions and Legendre polynomials, for $c = 0$ both the prolate Gauss-Lobatto and prolate-Lobatto quadratures are identical to Legendre Gauss-Lobatto quadrature.

The MATLAB programs `prolate_gauss_lobatto.m` and `prolate_lobatto.m` in Listings 3.3 and 3.4 provide implementations of these two schemes. Here, **Newton's method** [Kel95] is employed for the nonlinear equation solving and root-finding. For $c = 0$, the Legendre Gauss-Lobatto nodes are used as the initial guess. A numerical continuation procedure on parameter c is then employed to generate the initial guesses for subsequent values of c, stepping c up to the value for which the nodes are desired. Note that only *scalar* Newton iterations are required in the computation of the prolate-Lobatto nodes (each root can be computed independently of the others), as opposed to the more expensive *vector* Newton iterations necessary for computing the prolate Gauss-Lobatto nodes and weights.

While it may seem wasteful to compute the nodes for all the intermediate values of c, the procedure can still be performed efficiently. This is because, as shown in Figure 3.1, the nodes vary not only continuously but also relatively smoothly with c, so that large steps can be taken. The computational cost

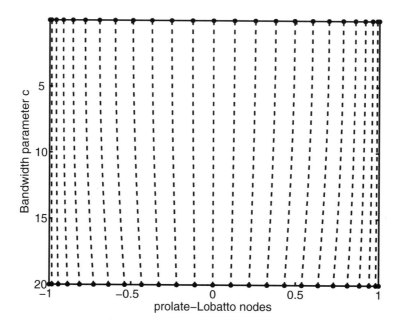

Figure 3.1: Computation of the $N = 25$ point prolate-Lobatto nodes by numerical continuation on the bandwidth parameter c. For $c = 0$, the prolate-Lobatto nodes are the Legendre Gauss-Lobatto nodes. Notice how the density of grid positions gradually increases in the center of the domain, as the highly non-uniform Legendre grid 'transforms' into the more uniform prolate grid.

of the continuation procedure can be further reduced by utilizing adaptive stepping: starting with a large step and decreasing it until divergence or duplication of roots is avoided.[2] The implementations can also be optimized by exploiting the fact that the nodes and weights are symmetric (see [Boy04, Boy05] for more details). An alternate approach, based on extrapolation for the efficient computation of the prolate Gauss-Lobatto nodes, has been given in [Boy05].

Figure 3.2 shows a comparison of the $N = 15$ point prolate-Lobatto grid (for $c = 15$) with the $N = 15$ point Chebyshev Gauss-Lobatto grid, in the 2-D interval $[-1, 1] \times [-1, 1]$. We can see that while both the tensor-product grids are non-uniform, the prolate grid is more uniform near the center of the domain, achieving better resolution therein than the Chebyshev grid.

---

[2]These conditions are easy to test because when Newton's method converges, it usually does so at a fast second-order exponential rate. In practice, only a few iterations are required for convergence up to machine precision.

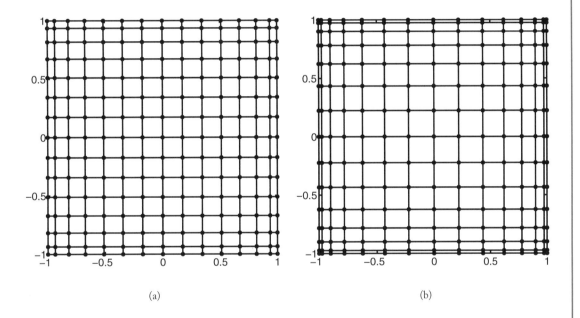

(a)   (b)

Figure 3.2: Comparison of the $N = 15$ point grids in the 2-D interval $[-1, 1] \times [-1, 1]$. (a) prolate-Lobatto grid (with $c = 15$) and (b) Chebyshev Gauss-Lobatto grid.

```
function [x,w] = prolate_gauss_lobatto (c,N,c_step,tol)

% N-point prolate Gauss-Lobatto nodes and associated weights for
% bandlimited function quadrature

Np = 2*N-2; M = 2*Np+30;
% Legendre Gauss-Lobatto nodes as initial guess for c = c_step
[x,w] = legendre_gauss_lobatto (N); v = [x(2:N-1); w];
% Continuation over parameter c
for c_i = c_step : c_step : c,
    % Expansion coefficients in Legendre basis
    [beta] = prolate_legendre_expansion (c_i,Np);
    % First Newton step
    x = [1; v(1:N-2); -1]; w = v(N-1:end);
    [P,D1P,D2P] = legendre_fn (x,M); Psi = P*beta;
    D1Psi = D1P*beta; F = Psi'*w-sqrt(2)*beta(1,:)';
    J = [D1Psi(2:N-1,:)'.*(ones(Np,1)*w(2:N-1)') Psi'];
    v_new = v - J\F; res = norm(v_new-v)/norm(v); v = v_new;
    % Newton iterations until convergence
    while res > tol
        x = [1; v(1:N-2); -1]; w = v(N-1:end);
        [P,D1P,D2P] = legendre_fn (x,M); Psi = P*beta;
        D1Psi = D1P*beta; F = Psi'*w-sqrt(2)*beta(1,:)';
        J = [D1Psi(2:N-1,:)'.*(ones(Np,1)*w(2:N-1)') Psi'];
        v_new = v - J\F; res = norm(v_new-v)/norm(v); v = v_new;
    end
end
x = [1; v(1:N-2); -1];          % Prolate Gauss-Lobatto nodes
w = v(N-1:end);                 % Prolate Gauss-Lobatto weights
```

Listing 3.3: `prolate_gauss_lobatto.m`

```
function [x,w] = prolate_lobatto (c,N,c_step,tol)

% N-point prolate-Lobatto nodes (roots of (1-x^2)*\psi_{N-1}'(x))
% and associated weights for bandlimited function quadrature

M = 2*N+30;
% Legendre Gauss-Lobatto nodes as initial guess for c = c_step
[x,w] = legendre_gauss_lobatto (N); x = x(2:(N-2-mod(N,2))/2+1);
% Continuation over parameter c
for c_i = c_step : c_step : c,
    % Expansion coefficients in Legendre basis
    [beta] = prolate_legendre_expansion (c_i,N);
    % First Newton step
    [P,D1P,D2P] = legendre_fn (x,M); D1Psi = D1P*beta(:,N);
    D2Psi = D2P*beta(:,N); x_new = x - D1Psi./D2Psi;
    res = norm(x_new-x)/norm(x); x = x_new;
    % Newton iterations until convergence
    while res > tol
        [P,D1P,D2P] = legendre_fn (x,M); D1Psi = D1P*beta(:,N);
        D2Psi = D2P*beta(:,N); x_new = x - D1Psi./D2Psi;
        res = norm(x_new-x)/norm(x); x = x_new;
    end
end
x = [1; x; zeros(mod(N,2),1);...
    -flipud(x); -1];                % Prolate-Lobatto nodes
[P,D1P,D2P] = legendre_fn (x,M); Psi = P*beta;
w = (Psi')\(sqrt(2)*beta(1,:)');    % Prolate-Lobatto weights
```

Listing 3.4: `prolate_lobatto.m`

CHAPTER 4

Applications

In this chapter, we present some example applications of the Gaussian quadrature methods discussed in the previous chapters.

4.1 INTEGRATION EXAMPLES

We consider the evaluation of the following integrals by Gaussian quadrature:

1. $\int_{-1}^{1} (1 + x^2)^{-1}\, dx = \pi/2$;

2. $\int_{-1}^{1} \cos(2\pi x)\, (1 - x^2)^{-1/2}\, dx = \pi J_0(2\pi)$;

3. $\int_{-1}^{1} \operatorname{sinc}(3x)\, dx = 2\operatorname{Si}(3\pi)/3\pi$;

4. $\int_{0}^{2\pi} \cos(\sin x)\, dx = 2\pi J_0(1)$,

where J denotes Bessel function of the first kind and Si denotes the sine integral function. The integrals are already in standard form so that no change of variables is required and the quadrature formulas can be used directly.[1] We select Legendre quadrature for evaluating the first integral, Chebyshev quadrature for the second, and prolate quadrature for the third. The fourth integral involves a periodic integrand, and we use Fourier Gaussian quadrature, which is based on trigonometric interpolants and equispaced nodes. This kind of quadrature is also called the **periodic trapezoid rule** and has nodes and weights given explicitly by

$$x_j = \frac{2j\pi}{N}, \quad w_j = \frac{2\pi}{N}, \text{ for } j = 0, \ldots, N - 1. \tag{4.1}$$

Figure 4.1 shows the accuracy of the polynomial Gaussian quadratures used to evaluate the first two integrals as a function of N. The quadratures are clearly spectrally accurate: as N increases, the relative error decreases exponentially until it reaches about 10^{-15}, after which, round-off error takes over. Note that, as expected, the Gauss-Lobatto quadratures are slightly less accurate than conventional Gaussian quadratures.

Figure 4.2 examines the accuracy of the prolate Gaussian quadratures used for evaluating the third integral. Figure 4.2(a) shows the accuracy of the quadratures as a function of bandwidth parameter c. The integrand $\operatorname{sinc}(3x) = \sin(3\pi x)/(3\pi x)$ is a bandlimited function with bandwidth 3π. This explains the optimum performance of the prolate Gauss-Lobatto and prolate-Lobatto quadratures near $c = 3\pi$ and $c = 3\pi/2$, respectively. With these choices of c, the prolate Gaussian quadratures are orders of magnitude more accurate than Legendre Gaussian quadrature (for $c = 0$, both prolate quadratures are identical to

---

[1] For integrals on the general (finite) interval $[a, b]$, the change of variables formula $\int_{a}^{b} f(x)\, dx = \frac{b-a}{2} \int_{-1}^{1} f(\frac{b-a}{2}t + \frac{b+a}{2})\, dt$ can be used. Similarly, integrals on an infinite interval can also be transformed, if needed, to integrals on a finite interval.

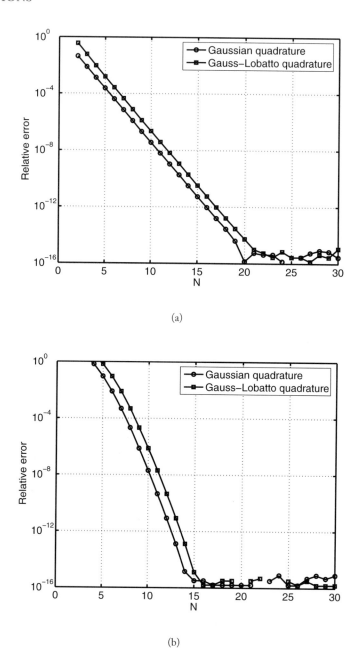

(a)

(b)

Figure 4.1: Accuracy of polynomial Gaussian quadratures: (a) Legendre quadrature for $\int_{-1}^{1}(1 + x^2)^{-1}\,dx$ and (b) Chebyshev quadrature for $\int_{-1}^{1}\cos(2\pi x)(1 - x^2)^{-1/2}\,dx$.

Legendre Gauss-Lobatto quadrature). Note, in addition, that with both schemes using the optimum c, the prolate Gauss-Lobatto quadrature yields a more accurate result than the prolate-Lobatto quadrature. This may be explained by the truncation limits in the error-bound formulas (3.22) and (3.36) for conventional (interior) prolate Gaussian quadratures (the end-point/Lobatto prolate Gaussian quadratures here could have similar error-bounds). Figure 4.2(b) shows the accuracy of the prolate Gaussian quadratures as a function of N (with c chosen near optimally: $c = 10$ is used for prolate Gauss-Lobatto quadrature and $c = 5$ is used for prolate-Lobatto quadrature). The results show spectral accuracy, with prolate Gauss-Lobatto quadrature being more accurate than prolate-Lobatto quadrature.

Figure 4.3 shows the accuracy of Gaussian quadratures for the fourth integral as a function of N. As seen from Figure 4.3(a), the periodic trapezoid rule is spectrally accurate, with a mere seven points yielding near double precision accuracy. It is interesting to compare the performance of Legendre Gaussian quadrature for the same integral, which, as shown in Figure 4.3(b), is also spectrally accurate, but requires about 25 points to achieve double precision accuracy. This example shows that for integrals involving smooth periodic integrands, the periodic trapezoid rule is very effective.

4.2   COMPUTATIONS INVOLVING CONTOUR INTEGRATION

We consider some examples of computing special integrals and functions using contour integration.

4.2.1   OSCILLATORY INTEGRAL

Consider the computation of:

$$I = \lim_{\varepsilon \to 0} \int_{\varepsilon}^{1} x^{-1} \cos(x^{-1} \log x) \, dx. \tag{4.2}$$

This problem appeared in the famous "Hundred-dollar, Hundred-digit Challenge Problems", a set of ten problems in numerical analysis published by Lloyd N. Trefethen in the January/February 2002 issue of *SIAM News*.

The integral can be evaluated in many ways, and its value $I = 0.323367431677778761399\ldots$ is known to more than 10,000 digits [BLWW04]. One approach is based on the use of **contour integration**, writing the integral as

$$I = \text{Re} \int_{C} z^{i/z-1} \, dz, \tag{4.3}$$

where C is any contour from 0 to 1 within which the integrand is analytic. Let C be a parameterized semi-circular curve of radius $r = 1/2$ and centered at $c = 1/2$: $z(\theta) = c + re^{i\theta}$, $\theta \in [0, \pi]$. The integrand is very smooth along this contour, and spectral accuracy can easily be achieved using a quadrature rule like Legendre Gaussian quadrature. This is implemented in the MATLAB program osc.m shown in Listing 4.1.

The output from this program is as follows:

```
N =   10, I_N = 0.32387119038233
N =   20, I_N = 0.32336284718615
N =   30, I_N = 0.32336740012902
```

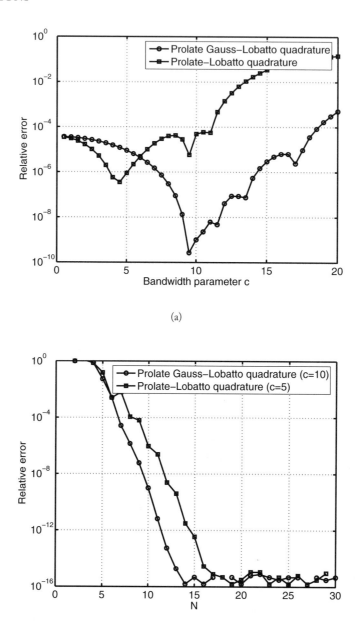

(a)

(b)

Figure 4.2: Accuracy of prolate Gaussian quadratures for $\int_{-1}^{1} \text{sinc}(3x)\, dx$: (a) as a function of c (with $N = 10$) and (b) as a function of N.

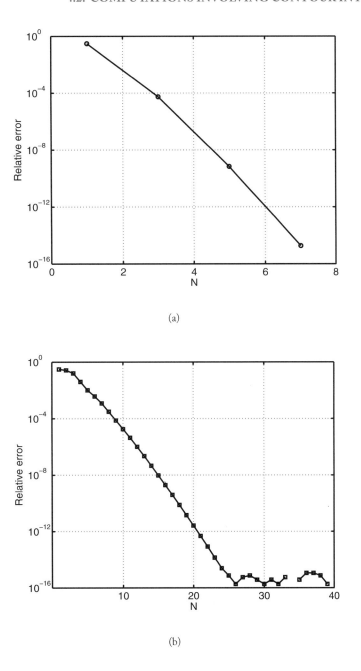

Figure 4.3: Accuracy of Gaussian quadratures for $\int_0^{2\pi} \cos(\sin x)\, dx$: (a) Fourier quadrature and (b) Legendre quadrature.

```
% Oscillatory integral via contour integration


I = 0.32336743167777876140;                % Exact integral to 20 digits
N_vec = [10:10:100];
err_vec = zeros(size(N_vec));
for n = 1 : length(N_vec),
    N = N_vec(n);
    % Legendre Gaussian quadrature
    [x,w] = legendre_gauss_quad (N);        % x goes from -1 to 1
    theta = pi/2*(x+1);                     % theta goes from 0 to pi
    c = 1/2; r = 1/2;                       % center and radius of contour
    z = c + r*exp(i*theta); dz = i*(z-c);   % contour (semi-circular arc)
    f = real((z.^(i./z-1)).*dz);            % integrand
    I_N = -pi/2*sum(w.*f);                  % integral
    fprintf('N = %3d, I_N = %15.14f\n',N,I_N);
    err_vec(n) = abs(I_N-I);
end
figure, semilogy(N_vec,err_vec,'o-');
xlabel('N'); ylabel('Absolute error');
```

Listing 4.1: osc.m

```
N =   40, I_N = 0.32336743065788
N =   50, I_N = 0.32336743166116
N =   60, I_N = 0.32336743167849
N =   70, I_N = 0.32336743167776
N =   80, I_N = 0.32336743167778
N =   90, I_N = 0.32336743167778
N =  100, I_N = 0.32336743167778
```

Figure 4.4 shows the accuracy in the computation of the oscillatory integral using contour integration with Legendre Gaussian quadrature as a function of N. We see that the result is spectrally accurate, with around 80 points yielding near double precision accuracy.

4.2.2   MATRIX FUNCTIONS

Consider the computation of the cube root of the matrix

$$A = \begin{bmatrix} 3 & 2 \\ 1 & 4 \end{bmatrix}. \tag{4.4}$$

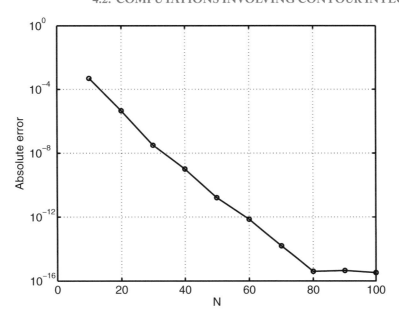

Figure 4.4: Accuracy in the computation of the oscillatory integral using contour integration with Legendre Gaussian quadrature.

The cube root is given by

$$B = A^{1/3} = \begin{bmatrix} \frac{1}{3}(2\sqrt[3]{2} + \sqrt[3]{5}) & -\frac{2}{3}(\sqrt[3]{2} - \sqrt[3]{5}) \\ \frac{1}{3}(-\sqrt[3]{2} + \sqrt[3]{5}) & \frac{1}{3}(\sqrt[3]{2} + 2\sqrt[3]{5}) \end{bmatrix}. \tag{4.5}$$

This problem can be solved with Newton's method, among other things, but here we describe an approach based on contour integration. Let f be an analytic function and A a square matrix. Then $f(A)$ can be defined by the contour integral [HHT08]

$$f(A) = \frac{1}{2\pi i} \int_\Gamma (zI - A)^{-1} f(z) \, dz, \tag{4.6}$$

where Γ is a closed contour enclosing the spectrum $\sigma(A)$. The MATLAB program `Afun.m`, shown in Listing 4.2, takes Γ as a parameterized circular contour of radius $r = 2.5$ and centered at $c = 3.5$: $z(\theta) = c + re^{i\theta}$, $\theta \in [0, 2\pi]$ and uses the trapezoidal rule to compute the resulting periodic integral with spectral accuracy.

The output from this program is:

```
N = 15, B = [1.40990083778652 0.30092673019030;
0.15046336509515 1.56036420288167]
N = 30, B = [1.40993943228961 0.30003677403587;
0.15001838701793 1.55995781930754]
```

```
% Cube root of the matrix A = [3 2; 1 4] via contour integration

A = [3 2; 1 4];                  % Matrix A
f = inline('z^(1/3)');           % Matrix function
N_vec = [15:15:90];
err_vec = zeros(size(N_vec));
for n = 1 : length(N_vec),
    N = N_vec(n);                % N
    % Trapezoidal rule
    theta = [0:N-1]*2*pi/N;      % theta goes from 0 to 2*pi
    c = 3.5; r = 2.5;            % center and radius of contour
    z = c + r*exp(i*theta);     % circular contour
    I = eye(2); B = zeros(2);
    for m = 1 : N,
        B = B + inv(z(m)*I-A)*f(z(m))*(z(m)-c);
    end
    B = real(B)/N;              % integral
    fprintf('N = %2d, B = [%15.14f %15.14f; %15.14f %15.14f]\n',...
        N,B(1,1),B(1,2),B(2,1),B(2,2));
    err_vec(n) = norm(B^3-A);   % Error
end
figure, semilogy(N_vec,err_vec,'o-');
xlabel('N'); ylabel('||B^3-A||');
```

Listing 4.2: Afun.m

```
N = 45, B = [1.40993934965273 0.30003659764187;
0.15001829882093 1.55995764847366]
N = 60, B = [1.40993934881845 0.30003659785641;
0.15001829892821 1.55995764774665]
N = 75, B = [1.40993934882217 0.30003659785454;
0.15001829892727 1.55995764774944]
N = 90, B = [1.40993934882215 0.30003659785455;
0.15001829892727 1.55995764774942]
```

Figure 4.5 shows the accuracy in the computation of the matrix cube root using contour integration with the periodic trapezoid rule as a function of N. The result is spectrally accurate, with around 90 points yielding near double precision accuracy.

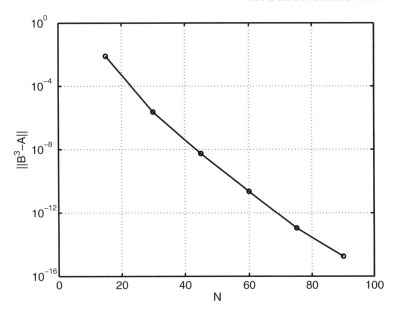

Figure 4.5: Accuracy in the computation of the matrix cube root using contour integration with the periodic trapezoid rule.

See [ST07] for a similar approach used for the computation of the Gamma function [AS65].

4.3 PSEUDOSPECTRAL METHODS

4.3.1 BACKGROUND

Pseudospectral methods form a powerful class of numerical methods for the accurate and efficient solution of ordinary and partial differential equations. Pseudospectral methods are based on the **Galerkin approach**, in which the problem is first discretized by using a convenient semi-discrete representation to approximate the true solution, and the solution of the resulting discretization is computed on this subspace. Provided that the discretization is consistent, the method yields a good approximation to the solution of the original problem.

Consider the representative equation [CHQZ88]

$$
\begin{aligned}
Lf &= g \text{ in } \Omega, \\
Bf &= 0 \text{ on } \partial\Omega_b,
\end{aligned}
\tag{4.7}
$$

where Ω is an open bounded domain in $\mathbb{R}^d$ with piecewise smooth boundary $\partial\Omega$, L denotes a differential operator (assumed linear and time-independent) in Ω, B is a linear boundary differential operator on

$\partial\Omega_b \subseteq \partial\Omega$, g is a known function, and f is the unknown function to be determined. Equation (4.7) constitutes a model boundary value problem, and it is assumed that the problem is well-posed.[2]

In Galerkin methods, we seek solutions of the form

$$f \approx f^N = \sum_{k \in J} \alpha_k\, \phi_k, \tag{4.8}$$

where J is a set of indices, ϕ_k are basis functions, and α_k are the expansion coefficients. The **Galerkin discretization** of (4.7) is defined by [CHQZ88]

$$\begin{aligned}(Lf^N, \phi_k) &= (g, \phi_k) \text{ for all } k \in J, \\ Bf^N &= 0 \text{ on } \partial\Omega_b,\end{aligned} \tag{4.9}$$

where the notation $(\cdot, \cdot)$ is used to denote inner-product. The Galerkin solution of (4.7) is computed by using (4.8) for f^N in (4.9) and solving for the unknown expansion coefficients α_k.

Spectral methods [Boy01, CHQZ88, For98, GO77, KS99, Tre96, Tre00] are Galerkin methods in which trigonometric and orthogonal polynomials—and more recently, prolate spheroidal wave functions (see [Boy04, BS05, CGH05, KLC05])—are utilized as basis functions. When the functions involved in the problem are smooth, thanks to the approximation theory of these basis functions, only a small number of terms are required in the expansion for the semi-discrete representation (4.8)—and through it, the Galerkin discretization—to be highly accurate. Spectral methods then generally yield *exponential* accuracy in the results: the decay of the error in the solution is faster than any finite power of $1/N$ as $N \to \infty$ (where N is the unknown number).

The **pseudospectral discretization** of the model boundary value problem (4.7) is defined by [CHQZ88]

$$\boxed{\begin{aligned}Lf^N(x_k) = g(x_k) &\quad \text{for all } k \in J_e, \\ Bf^N(x_k) = 0 &\quad \text{for all } k \in J_b,\end{aligned}} \tag{4.10}$$

where x_k, $k \in J$, are a set of 'collocation' positions in the domain Ω or on its boundary $\partial\Omega$ (J being the set of indices), and J_e and J_b are two disjoint subsets of J such that if $k \in J_e$, the x_k's are in Ω, and if $k \in J_b$, the x_k's are on the part $\partial\Omega_b$ of the boundary where the boundary conditions are specified. The pseudospectral solution of (4.7) is computed by using (4.8) for f^N in (4.10) and solving for the unknown expansion coefficients α_k.

The pseudospectral (or collocation) discretization (4.10) can be derived from the Galerkin spectral discretization (4.9) by approximating the continuous inner-product $(\cdot, \cdot)$ with a discrete inner-product $(\cdot, \cdot)_N$, defined as

$$(u, v)_N \triangleq \sum_{k \in J} u(x_k)\, \overline{v(x_k)}\, w_k. \tag{4.11}$$

---

[2]Precise conditions for the well-posedness of this problem can be found in [CHQZ88].

The pseudospectral discretization (4.10) can then be equivalently written as

$$
\begin{aligned}
(Lf^N, \phi_k)_N &= (g, \phi_k)_N & \text{for all } k \in J_e, \\
Bf^N(x_k) &= 0 & \text{for all } k \in J_b.
\end{aligned}
\tag{4.12}
$$

If the nodes $\{x_k\}$ and weights $\{w_k\}$ are such that the continuous inner-product is well approximated by the discrete inner-product (as it is, if the nodes and weights are chosen according to the Gaussian quadrature rules), the pseudospectral discretization (4.10) provides a good approximation to the spectral Galerkin discretization (4.9). In particular, the form of the integrals suggests that a good choice would be to utilize the nodes and weights of the N-point Gaussian (or Gauss-Lobatto) quadrature associated with the basis functions ϕ_k. Notice that the pseudospectral discretization only utilizes the nodes and does not require the weights of the quadrature.

In the pseudospectral discretization (4.10), a semi-discrete representation in terms of trigonometric or orthogonal polynomials and related functions is used to approximate the unknown function f and the original equation (4.7) is required to be satisfied exactly at the nodes of the Gaussian quadrature associated with the basis functions. The method is therefore also known as **pseudospectral collocation**.

4.3.2   LEGENDRE PSEUDOSPECTRAL METHOD FOR THE FRESNEL INTEGRALS

We consider the problem of computing the Fresnel integrals [AS65]

$$
C(x) = \int_0^x \cos\left(\frac{\pi}{2}t^2\right) dt, \text{ and}
\tag{4.13a}
$$

$$
S(x) = \int_0^x \sin\left(\frac{\pi}{2}t^2\right) dt,
\tag{4.13b}
$$

which are encountered frequently in diffraction theory. Writing the integrals as ordinary differential equation problems, we have

$$
C'(x) = \cos\left(\frac{\pi}{2}x^2\right), \ C(0) = 0, \ x > 0, \text{ and}
\tag{4.14a}
$$

$$
S'(x) = \sin\left(\frac{\pi}{2}x^2\right), \ S(0) = 0, \ x > 0.
\tag{4.14b}
$$

We now solve (4.14) using the Legendre pseudospectral method. The MATLAB program `legendre_ps_fresnel.m` in Listing 4.3 implements the Legendre pseudospectral method for this problem. Figure 4.6 shows a comparison of the output of this program (with $N = 30$ unknowns in the Legendre pseudospectral method) to the plots of the Fresnel integrals in Figure 7.5 of [AS65]. Figure 4.7 shows the relative $\mathbf{L}^2$ error in the Fresnel integrals from the Legendre pseudospectral result. Spectral accuracy is evident, with $N \approx 50$ sufficient to obtain near double precision accuracy. On a 2.8 GHz Intel Pentium 4 processor, the program runs in less than 0.2 seconds.

In [WR00], a MATLAB based software suite is developed for pseudospectral differentiation and interpolation, and several applications are given involving the solution of differential equations for the computation of the complimentary error function [AS65] and eigenvalue and boundary value problems. In [Tre00], the realization of Fourier and Chebyshev based spectral methods in MATLAB is discussed

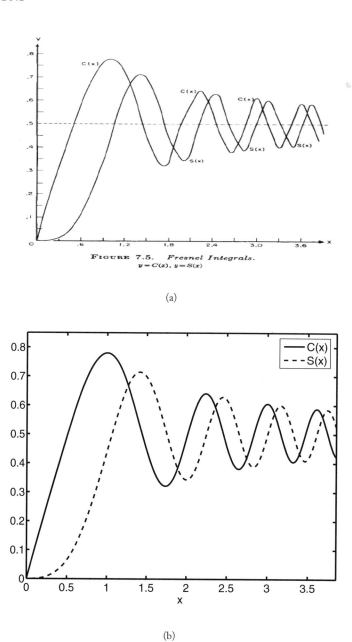

(a)

(b)

Figure 4.6: Fresnel integrals: (a) Figure 7.5 from Abramowitz and Stegun (1965) and (b) Legendre pseudospectral method with $N = 30$ (output from `legendre_ps_fresnel.m` in Listing 4.3).

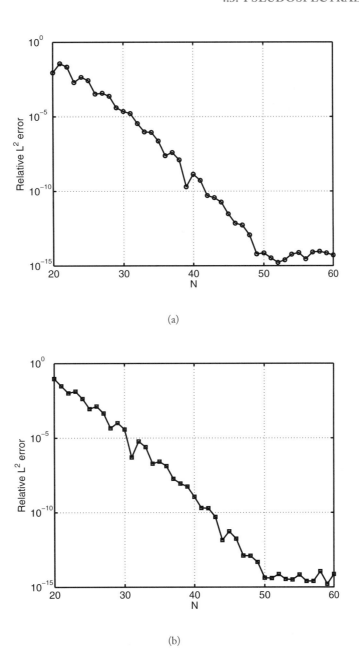

(a)

(b)

Figure 4.7: Relative error in the Fresnel integrals from the Legendre pseudospectral method: (a) $C(x)$ and (b) $S(x)$.

```
% Legendre pseudospectral method for the Fresnel integrals

N = 30;                                % Unknown number
xm = 3.85;                             % Domain is in [0,xm]


[x,w] = legendre_gauss_lobatto (N);    % Legendre Gauss-Lobatto pts
[P,D1P,D2P] = legendre_fn (x,N-1);     % Legendre polynomials
x = xm*(x+1)/2; L = 2/xm*D1P;          % Domain and system mapping
g1 = cos(pi/2*x.^2); g2 = sin(pi/2*x.^2);% Right-hand sides
L(N,:) = P(N,:); g1(N) = 0; g2(N) = 0;   % Boundary conditions
alpha1 = L\g1; alpha2 = L\g2;          % Expansion coefficients


x_i = [1:-0.01:-1]'; x = xm*(x_i+1)/2; % Interpolation positions
[P,D1P,D2P] = legendre_fn (x_i,N-1);   % Legendre polynomials
f1 = P*alpha1; f2 = P*alpha2;          % Fresnel integrals
figure, plot(x,f1); xlabel('x');
hold on, plot(x,f2,'--'); legend('C(x)','S(x)');
axis([0 xm 0 0.85]);
```

Listing 4.3: `legendre_ps_fresnel.m`

and illustrated through several examples. In [CGH05, KLC05], the prolate pseudospectral method is developed and applied to solve PDEs for wave propagation and electromagnetics problems.

APPENDIX A

Links to Mathematical Software

Below are links to some useful mathematical software available on the Internet.

- **NETLIB** - A freely available software repository for numerical and scientific computing.
  http://www.netlib.org/

- **LAPACK** - A software package which provides routines for solving linear algebra problems.
  http://www.netlib.org/lapack/

- **GSL** - A freely available library of numerical and scientific computing routines for C and C++ programmers.
  http://www.gnu.org/software/gsl/

- **CHEBFUN** - A software system in MATLAB for numerical computing with continuous or piecewise-continuous functions.
  http://www2.maths.ox.ac.uk/chebfun/

- **OCTAVE** - A MATLAB-like high-level interpreted language and freely available interactive software package for numerical computations.
  http://www.gnu.org/software/octave/

Bibliography

[ABB+99] E. Anderson, Z. Bai, C. Bischof, S. Blackford, J. Demmel, J. Dongarra, J. Du Croz, A. Green-baum, S. Hammarling, A. McKenney, and D. Sorensen. *LAPACK Users' Guide.* Society for Industrial and Applied Mathematics, Philadelphia, PA, third edition, 1999. Cited on page(s) 9, 24

[AS65] M. Abramowitz and I. A. Stegun, editors. *Handbook of Mathematical Functions, with Formulas, Graphs, and Mathematical Tables.* Dover, New York, 1965. Cited on page(s) 25, 45, 47

[Atk89] K. E. Atkinson. *An Introduction to Numerical Analysis.* Wiley, second edition, 1989. Cited on page(s) 1, 23

[BLWW04] F. Bornemann, D. Laurie, S. Wagon, and J. Waldvogel. *The SIAM 100-Digit Challenge: A Study in High-Accuracy Numerical Computing.* SIAM, 2004. Cited on page(s) 39

[Bou47] C. J. Bouwkamp. On spheroidal wave functions of order zero. *Journal of Mathematics and Physics,* 26:79–92, 1947. Cited on page(s) 8, 9

[Boy01] J. P. Boyd. *Chebyshev and Fourier Spectral Methods.* Dover, New York, second edition, 2001. Cited on page(s) 46

[Boy04] J. P. Boyd. Prolate spheroidal wavefunctions as an alternative to Chebyshev and Legendre polynomials for spectral element and pseudospectral algorithms. *Journal of Computational Physics,* 199:688–716, 2004. DOI: 10.1016/j.jcp.2004.03.010 Cited on page(s) 9, 16, 19, 31, 32, 46

[Boy05] J. P. Boyd. Algorithm 840: Computation of grid points, quadrature weights and derivatives for spectral element methods using Prolate spheroidal wave functions - Prolate elements. *ACM Transactions on Mathematical Software,* 31:149–165, 2005. DOI: 10.1145/1055531.1055538 Cited on page(s) 9, 31, 32

[BS05] G. Beylkin and K. Sandberg. Wave propagation using bases for bandlimited functions. *Wave Motion,* 41:263–291, 2005. DOI: 10.1016/j.wavemoti.2004.05.008 Cited on page(s) 19, 46

[CC60] C. W. Clenshaw and A. R. Curtis. A method for numerical integration on an automatic computer. *Numerische Mathematik,* 2:197–205, 1960. DOI: 10.1007/BF01386223 Cited on page(s) 23

[CGH05] Q.-Y. Chen, D. Gottlieb, and J. S. Hesthaven. Spectral methods based on prolate spheroidal wave functions for hyperbolic PDEs. *SIAM Journal on Numerical Analysis,* 43:1912–1933, 2005. DOI: 10.1137/S0036142903432425 Cited on page(s) 31, 46, 50

[CH53]   R. Courant and D. Hilbert. *Methods of Mathematical Physics*, volume 1. Wiley Interscience, New York, NY, 1953. Cited on page(s) 5

[CHQZ88]   C. Canuto, M. Y. Hussaini, A. Quarteroni, and T. A. Zang. *Spectral Methods in Fluid Dynamics*. Springer-Verlag, Berlin, 1988. Cited on page(s) 13, 45, 46

[CRY99]   H. Cheng, V. Rokhlin, and N. Yarvin. Nonlinear optimization, quadrature, and interpolation. *SIAM Journal on Optimization*, 9:901–923, 1999. DOI: 10.1137/S1052623498349796 Cited on page(s) 8, 21

[For98]   B. Fornberg. *A Practical Guide to Pseudospectral Methods*. Cambridge, 1998. Cited on page(s) 46

[GL96]   G. H. Golub and C. F. Van Loan. *Matrix Computations*. Johns Hopkins University Press, Baltimore, third edition, 1996. Cited on page(s) 9, 24

[GO77]   D. Gottlieb and S. A. Orszag. *Numerical Analysis of Spectral Methods: Theory and Applications*. SIAM, Philadelphia, 1977. Cited on page(s) 13, 46

[Gol73]   G. H. Golub. Some modified matrix eigenvalue problems. *SIAM Review*, 15:318–334, 1973. DOI: 10.1137/1015032 Cited on page(s) 24, 29

[GS97]   D. Gottlieb and C.-W. Shu. On the Gibbs phenomenon and its resolution. *SIAM Review*, 39:644–668, 1997. DOI: 10.1137/S0036144596301390 Cited on page(s) 13

[GSSV92]   D. Gottlieb, C.-W. Shu, A. Solomonoff, and H. Vandeven. On the Gibbs phenomenon I: Recovering exponential accuracy from the Fourier partial sum of a non-periodic analytical function. Technical report, Institute for Computer Applications in Science and Engineering, 1992. Cited on page(s) 13

[GW69]   G. H. Golub and J. H. Welsch. Calculation of Gauss quadrature rules. *Mathematics of Computation*, 23:221–230+s1–s10, 1969. DOI: 10.1090/S0025-5718-69-99647-1 Cited on page(s) 24, 29

[HHT08]   N. Hale, N. J. Higham, and L. N. Trefethen. Computing A^α, $\log(A)$, and related matrix functions by contour integrals. *SIAM Journal on Numerical Analysis*, 46:2505–2523, 2008. DOI: 10.1137/070700607 Cited on page(s) 43

[Kel95]   C. T. Kelly. *Iterative Methods for Linear and Nonlinear Equations*. SIAM, Philadelphia, 1995. Cited on page(s) 31

[KLC05]   N. Kovvali, W. Lin, and L. Carin. Pseudospectral method based on prolate spheroidal wave functions for frequency-domain electromagnetic simulations. *IEEE Transactions on Antennas and Propagation*, 53:3990–4000, 2005. DOI: 10.1109/TAP.2005.859899 Cited on page(s) 31, 46, 50

[KS99]   G. E. Karniadakis and S. J. Sherwin. *Spectral/hp Element Methods for CFD*. Oxford University Press, Oxford, UK, 1999. Cited on page(s) 46

[LP61]   H. J. Landau and H. O. Pollak. Prolate spheroidal wave functions, Fourier analysis and uncertainty-II. *The Bell System Technical Journal*, 40:65–83, 1961. Cited on page(s) 8

[MRW96] J. Ma, V. Rokhlin, and S. Wandzura. Generalized Gaussian quadratures rules for systems of arbitrary functions. *SIAM Journal on Numerical Analysis*, 33:971–996, 1996. DOI: 10.1137/0733048 Cited on page(s) 8, 21

[Sle83] D. Slepian. Some comments on Fourier analysis, uncertainty and modeling. *SIAM Review*, 25:379–393, 1983. DOI: 10.1137/1025078 Cited on page(s) 8

[SP61]   D. Slepian and H. O. Pollak. Prolate spheroidal wave functions, Fourier analysis and uncertainty-I. *The Bell System Technical Journal*, 40:43–63, 1961. Cited on page(s) 8

[ST07]   T. Schmelzer and L. N. Trefethen. Computing the Gamma function using contour integrals and rational approximations. *SIAM Journal on Numerical Analysis*, 45:558–571, 2007. DOI: 10.1137/050646342 Cited on page(s) 45

[Sze75] G. Szegő. *Orthogonal Polynomials*. American Mathematical Society, Providence, RI, fourth edition, 1975. Cited on page(s) 6

[TB97]   L. N. Trefethen and D. Bau. *Numerical Linear Algebra*. SIAM, Philadelphia, 1997. Cited on page(s) 6, 9, 24, 29

[Tre96] L. N. Trefethen. Finite difference and spectral methods for ordinary and partial differential equations. unpublished text, available online at `http://web.comlab.ox.ac.uk/oucl/work/nick.trefethen/pdetext.html`, 1996. Cited on page(s) 46

[Tre00] L. N. Trefethen. *Spectral Methods in MATLAB*. SIAM, Philadelphia, 2000. Cited on page(s) 46, 47

[Tre08] L. N. Trefethen. Is Gauss quadrature better than Clenshaw-Curtis? *SIAM Review*, 50:67–87, 2008. DOI: 10.1137/060659831 Cited on page(s) 23

[WR00]   J. A. C. Weideman and S. C. Reddy. A MATLAB differentiation matrix suite. *ACM Transactions on Mathematical Software*, 26:465–519, 2000. DOI: 10.1145/365723.365727 Cited on page(s) 47

[XRY01] H. Xiao, V. Rokhlin, and N. Yarvin. Prolate spheroidal wavefunctions, quadrature and interpolation. *Inverse Problems*, 17:805–838, 2001. DOI: 10.1088/0266-5611/17/4/315 Cited on page(s) 8, 9, 14, 21, 26, 27, 28, 31

[YR98]   N. Yarvin and V. Rokhlin. Generalized Gaussian quadratures and singular value decompositions of integral operators. *SIAM Journal on Scientific Computing*, 20:699–718, 1998. DOI: 10.1137/S1064827596310779 Cited on page(s) 8, 21

Author's Biography

NARAYAN KOVVALI

Narayan Kovvali received the B. Tech. degree in electrical engineering from the Indian Institute of Technology, Kharagpur, India, in 2000, and the M.S. and Ph.D. degrees in electrical engineering from Duke University, Durham, North Carolina, in 2002 and 2005, respectively. In 2006, he joined the Department of Electrical Engineering at Arizona State University, Tempe, Arizona as an Assistant Research Scientist. He currently holds the position of Assistant Research Professor in the School of Electrical, Computer, and Energy Engineering at Arizona State University. His research interests include statistical signal processing, detection, estimation, stochastic filtering and tracking, Bayesian data analysis, multi-sensor data fusion, Monte Carlo methods, and scientific computing. Dr. Kovvali is a Senior Member of the IEEE.

Printed in the United States
by Baker & Taylor Publisher Services